THERE IS LIFE ON MARS

by the same author

LIFE AND THE UNIVERSE

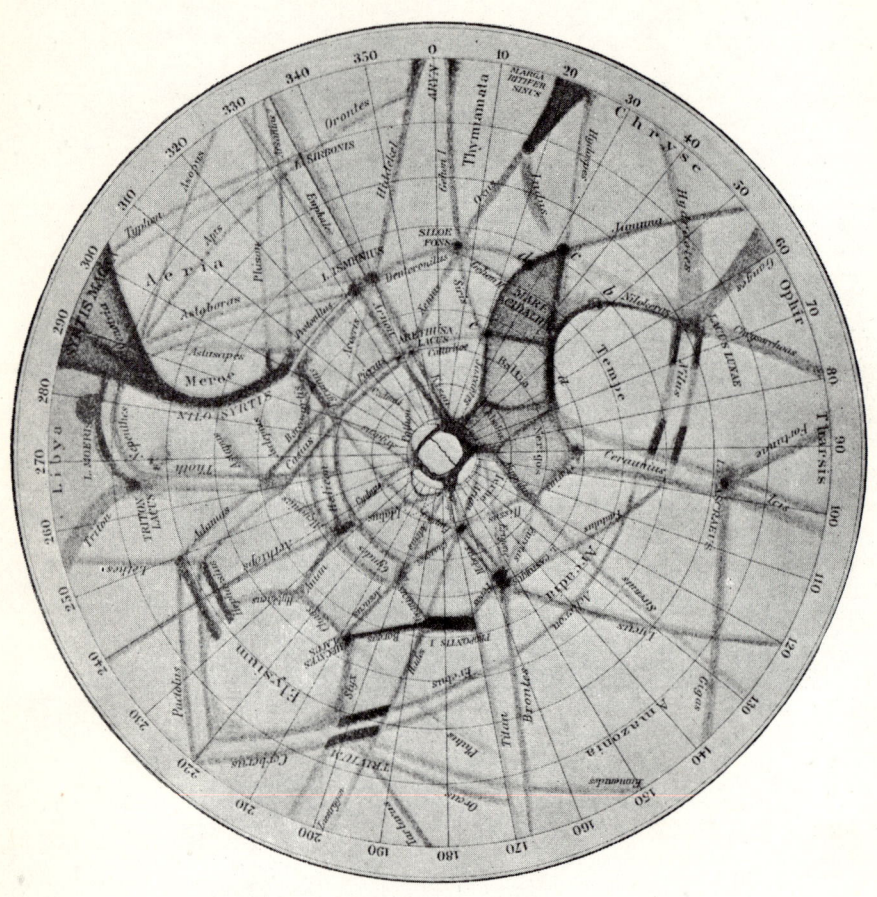

A chart of the Northern Hemisphere of Mars,
by Schiaparelli, 1888

Reproduced by courtesy of the Royal Astronomical Society

There IS Life On MARS

by

THE EARL NELSON
F.R.A.S., F.R.S.A., F.R.G.S., F.Z.S.

The Citadel Press
New York

Copyright 1955 by Albert Francis Joseph Horatio Nelson · First American edition 1956 · Library of Congress Catalog Card Number 56-11882 · Manufactured in the United States of America

To
the memory of
Dr. EDWIN HUBBLE
formerly Chairman of the Research Committee
for the Mount Wilson and Palomar Observatories
who encouraged me to go ahead with
the writing of this book

Contents

Foreword   11

I
The Universe Around Us—
The Carbon Atom—Cosmic Energy—Life   13

II
The Substance of Life—
Martian Conditions—Life on Mars   25

III
What is Life?   37

IV
Factors in Planetary Temperature Variations—
The Terrestial Glaciations   46

V
The Velocity of Escape—
Planetary Atmospheres   53

VI
The Canals of Mars—
Possible Forms of Martian Life—Martian Vegetation—Flying Saucers   58

VII
The Planet Mars   77

VIII
The Great Mystery of the Martian Canals   84

IX
The Planet Venus—
Possibility of Life There—Possible Life on the Moon   93

X
Life on Other Worlds—
On Mars—On Venus   103

XI
Inter-Planetary Travel
and Space Ships of the Future   109

XII
Landing on the Moon—
On Mars—The Lunar Base—The Martian Base   121

Notes   131

Bibliography   141

Index   147

Illustrations

A chart of the Northern Hemisphere of Mars, by Schiaparelli, 1888	*frontispiece*
Mars photographed in blue and red light (*Hale*)	*facing page* 64
Mars photographed in ultra-violet and infra-red light (*Wright*)	65
Mars, showing one of its polar snow caps (*Hale*)	65
The Moon from Eratosthenes to Plato	80
The Nebula in Andromeda	81

Foreword

LONG AND careful study of the information accumulated over many years by astronomers and scientists of various countries, and of recent observations and discoveries made, has led me to the conclusion that some form of life definitely exists on Mars.

It may be nothing more than a low form of vegetation of the lichen type, or possibly some kind of plant life peculiar to Mars, the nature of which we can at present only guess at.

The existence of animal life seems improbable in the light of our present knowledge, although the possibility cannot be entirely ruled out.

Most people, when they speak of life on Mars, picture some sort of intelligent life, creatures something like ourselves.

It is most unlikely that intelligent life exists there, although even this is not an impossibility. The popular idea of little Martian men, miniature replicas of ourselves, however, is pure moonshine.

Even if intelligent creatures did exist, owing to the very different physical conditions on Mars, they would probably bear no remote resemblance to us, except in having large and efficient brains, and an upright stance.

In 1956 Mars will be nearer to the Earth than it has been for many years, and will, in fact, make its closest possible approach. It will then be distant about thirty-five million miles.

This comparative proximity may afford us exceptionally favourable opportunities for observation, and perhaps lead to discoveries that will throw new light on conditions prevailing at the surface of the planet.

It will be appreciated, that in a work of this kind there is necessarily a good deal of theory, of speculation. For instance, in Chapter IV, we do not know definitely the cause of temperature variations that occur on the Earth. The different possibilities have been touched upon, and one or more of the factors involved may be responsible for the onset of the Ice Ages or Glaciations.

So far as possible, I have tried to draw something like a line between the speculative and the factual by using a form of words, where the question is in doubt, that is not too definite: " It may be ", " possibly ", " perhaps ", and so on.

That I stand to be shot at for calling this book by such a title, goes without saying. There are many scientists and astronomers, who, for one reason or another, have made up their minds that there is no life anywhere else but here, and some, even, who appear to harbour an active dislike of the very idea of life on Mars, or any other planet. I am sorry if I have offended them.

I have already been taken to task for some of my remarks in the chapter on inter-planetary travel, particularly in regard to the subject of space navigation.

Now, I do not pretend to be an authority on this subject, and the views I have given are those of people who are, in so far as anyone can claim to be, an authority on this science of the future, still in its earliest infancy.

CHAPTER I

The Universe Around Us

The Carbon Atom—Cosmic Energy—Life

As most people know, the solar system consists of a central star, the Sun, round which circle, at varying distances outwards, nine planets, Mercury, Venus, Earth, Mars, Jupiter, Saturn, Uranus, Neptune and Pluto. Between Mars and Jupiter there occurs what is known as the Zone of Asteroids, a vast number of planetoids or minor planets, the largest of which, Ceres, has a diameter of 485 miles. Observation seems to show that most of them are probably huge rocks, of irregular shape, flying round in space. From this it is inferred that they may be the fragments of a planet which exploded. Another theory is that they have resulted from collisions between small planets.

The Sun is one of at least a hundred thousand million stars belonging to what is known as the Galaxy, and not a very large or important one at that. That part of the Galaxy which we can actually see is the Milky Way. It forms perhaps a hundredth part of the whole. The remainder is hidden from us by clouds of obscuring matter, interstellar gas and dust. It is from this dust that the stars are formed, and it is thought that the total amount of matter in the Galaxy not yet formed into stars is about equal in quantity to that of all the existing stars.

The Galaxy is formed like a gigantic cartwheel with

the hub in the centre, and is rotating slowly under its own gravitational attraction. The diameter of this great wheel is about a hundred thousand light years, and the distance of the Sun from the hub or centre is about thirty thousand light years. A light year is the distance that light, moving at a speed of 186,283 miles per second, travels in one year (nearly six million million miles). Light traverses the ninety-three million miles between the Sun and the Earth in eight and one-third minutes. By comparison, it takes more than four years to travel from the next nearest star to our Sun, a distance of twenty-five million million miles. If we want to get some idea of what our Galaxy would look like viewed from a great distance away in space, we have only to glance at the photograph of the great Nebula in Andromeda which is generally considered to be the twin of our Galaxy or nebula. The Sun travels round the centre of gravity of the Galaxy together with the planets at a speed of nearly a million miles an hour. Even so, it takes about two hundred million years to make one round trip.

Within range of the 200 inch telescope at Mount Palomar in California, there are more than a hundred million similar galaxies which are spoken of as the *extra galactic nebulæ*. The farthest of these are at a distance of about one thousand million light years. When we look at a photograph of these most distant galaxies we are seeing them, not as they are to-day, but as they were a thousand million years ago. The Andromeda Galaxy is one of the nearest, being only about one million, four hundred thousand light years distant from us. If we had a telescope with twice the power of the one at Mount Palomar, we should no doubt be able to see other galaxies twice as far away as those we can see now. What lies beyond all that?

It seems probable that the galaxies spread out in all directions in space, to infinity, in a Universe that had no beginning and will have no end, and that is without limits. Even if we confine ourselves to the hundred million or more galaxies that are within range of the Palomar telescope, or to our own Galaxy with its hundred thousand million suns, it seems unlikely, to say the least, that our Sun is the only one with a retinue of planets on which life exists. If that were so, we can only assume that its existence here is due to some peculiar accident, a most unlikely supposition. As it happens, we now have evidence that two of the nearest stars, 61 *Cygni* and 70 *Ophiuchi*, have planetary bodies revolving round them.

It is understandable that long ago, when people believed that the Earth was the fixed and immutable centre of the Universe, they should have held the opinion that no life could exist anywhere else but here. Once it was proved, however, that the Earth is only one of the smaller planets, circling round a very average and unimportant star, the sun, it was inevitable that doubts should arise about its being the only home of life in the Universe.

One of the tenets of the theory of relativity is that matter behaves to-day, in the stars or on the planets, in exactly the same way as it behaved in distant stars or planets thousands of millions of years ago. In other words, matter behaves in the same way at all times and in all parts of the Universe. That is, I think, a strong argument in favour of life occurring wherever conditions for its existence are favourable. Another argument in favour of such a possibility is the peculiarity of the carbon atom.

THERE IS LIFE ON MARS

Living matter, as we know, is built up of quite ordinary atoms, but it consists mainly of atoms having the capacity to form molecules containing thousands, or tens of thousands, of atoms. Most atoms lack this property.

The atoms of hydrogen and oxygen will combine to form molecules of hydrogen (H_2 or H_3), of oxygen or ozone (O_2 of O_3), or of water (H_2O), or hydrogen peroxide (H_2O_2). Not one of these compounds, however, contains more than four atoms. Even the addition of nitrogen does not materially alter conditions. The compounds of hydrogen, oxygen and nitrogen all contain only a few atoms. But if carbon is added, something very extraordinary happens. The atoms of hydrogen, oxygen, nitrogen and *carbon* combine to form molecules containing tens of thousands of atoms. These are the kind of molecules of which living bodies are mainly built up. It looks as if life exists in the Universe only because the carbon atom possesses these extraordinary properties.

The carbon atom has six electrons revolving round the central nucleus and it seems to differ from the atoms of boron and nitrogen, which are nearest to it in the table of chemical elements, merely in having one electron more than the atom of boron, and one less than that of nitrogen. This difference, slight as it seems, would appear accountable for the existence of living matter. We do not know why the six-electron atom should possess these extraordinary properties. The reason, no doubt, lies somewhere in the fundamental laws of nature, and one day we may fathom it.

Similar cases of atoms having special properties are those connected with the phenomena of magnetism and radio-activity. Atoms having 26 to 28 electrons have the properties of magnetism, those with 83 to 92 electrons that of radio-activity. Again, we do not know why.

Apart from carbon, there is one other element that possesses the ability to build up very complex molecules, *silicon*. The compound molecules which have carbon as a basis are, however, both more numerous and more complex than those having silicon as a basis. It is not impossible that on some other planets a form of life might be based on the silicon atom. It would not be life as we understand it, and creatures built up of cells having a silicon basis would be able to live in temperatures that would prove fatal to us. They could inhabit worlds that would be impossible for us to exist on.

Ninety-two elements are known to exist and everything in the Universe, living and non-living, is built up of the atoms of these elements. The immense variety of substances existing is accounted for by the almost unlimited variety of ways in which the atoms of this comparatively limited number of elements can be combined. Since the atoms found everywhere in the Universe obey the same chemical laws, it is clear that similar chemical compounds can exist anywhere under similar physical conditions. Most elements are comparatively rare. Over ninety per cent. of the Earth's crust is built up of the elements of oxygen, silicon, magnesium and iron, while only about a thousandth part contains the fifty less common elements. Each element has its own particular kind of atom, which differs from those of other elements. A molecule is the basic unit of any chemical compound and the smallest particle of the substance that can exist separately. If, for example, a molecule of water is broken up, we do not get less water, but no water at all. Instead we have two atoms of hydrogen and one of oxygen, hydrogen and oxygen being the elements of water.

These are important facts to bear in mind when we come to consider the kind of life that may exist on any particular planet, or the conditions necessary for life to exist at all. There can be little doubt that life of some sort exists on countless planets throughout the Universe, and there are probably vast numbers on which we could live just as well as we can here, and on which there may be life very similar to that we know. There are probably vast numbers of planets on which intelligent life has come into being just as it has here. We may never be able to find proof of this but it is reasonable assumption, seeing that the number of stars in the Universe is perhaps without limit, or even taking into account only those now within range of the Palomar telescope, which must be counted in millions of millions. What we are really concerned with, however, is the number of planets within our own particular solar system which may have some sort of life existing on them.

In the light of our present knowledge it would seem that there are only three planets in the solar system on which conditions are such that life might be possible: the Earth, Venus and Mars. Mercury, owing to its nearness to the Sun is far too hot, while the remainder are far too cold. When it comes to the exciting prospect of discovering *intelligent life* on other worlds, we have to take into consideration the fact that the time during which such a form of life may endure, forms only an infinitesimal fraction of the total life of a planet.

The Earth, for example, is some three thousand, five hundred million years old and for at least a thousand million years—possibly for nearly twice that period—

life in some form has existed here. By comparison, intelligent life has had a very brief existence. It is only during the last million years that anything which could be described as *human* has evolved from the earlier ape-men, the ancestors of the human race, who, in turn were descended from pre-historic apes. This is not theory or surmise but the proven facts of modern investigation.

So far as we know, life can only exist within narrow limits of temperature, that is, on planets which are neither too far away from, nor too near to a central sun. It is unlikely, therefore, that in any solar system life would attain the same stage of development on more than one planet at any one particular period of time. Over a long period, life might conceivably more or less die out on one planet, while coming to full flower in another more distant one, owing to a gradual increase of temperature in the central sun. The opposite might occur through a gradual decrease of temperature.

As we know to-day, everything in the Universe is built up of atoms and the atom is electric. Life may, therefore, be defined as an electro-chemical phenomenon.

This ties up with the theory of the continuous creation of matter. Presumably the created matter is in the form of atoms of hydrogen, since everything in the Universe originates as hydrogen. We may ask, where does the newly created matter, namely atoms of hydrogen, come from?

I think it is possible that energy of extremely short wave length, shorter than anything we know at present, might build up into atoms of hydrogen. If we ask how the energy originates, I would be inclined to answer that it does not originate. Like the Universe itself, it was always there and always will be. In other words, it is eternal and indestructible.

The Universe, with everything in it, exists in two forms—*energy* and *matter*. Energy cannot exist without matter, nor matter without energy, they being merely different forms of the same thing. We do not know if *matter-energy* ever had a beginning or if it will ever have an end. The probability is that it is eternal, a fact difficult for us to grasp living, as we do, on a finite speck of cosmic dust, the Earth; which will endure for a time and then cease to exist, at any rate in its present form, as a home of life. Time itself may have no real existence outside our limited experience. Time, for the Universe, being possibly one continuous present.

The cosmic energy of the Universe, is locked up in the nuclei of all atoms and, so far, there are only two known methods of releasing it; by *fission*, namely the splitting of the nuclei of the heaviest chemical elements into two unequal fragments consisting of the nuclei of two lighter elements, or by *fusion*, that is fusing two nuclei of the lightest elements into one nucleus of a heavier element. Fusion is now considered to be the means by which the radiant energy given out by the Sun and other stars is maintained over thousands of millions of years. It is a process by which four atoms of *hydrogen* are converted into one atom of *helium*. It is as a result of this process taking place in the Sun and other stars that life was brought into existence on the Earth, and probably on countless other planets throughout the Universe, and it is by the same process, the gradual heating up of the Sun as more and more hydrogen is converted into helium, that life on Earth, and eventually the Earth itself, will be destroyed. No one need worry unduly about this eventuality; however.

So far as we can tell, it is thousands of millions of years distant and long before it comes about, the human race may have disappeared like the extinct creatures of a past era.

The atomic bomb is dependent on fission for its effects, while the much more powerful hydrogen bomb is based on the process of fusion. It is possible that even fission and fusion may not be the end of the story. With these, only a small fraction of the mass of the protons and neutrons in the nuclei of the elements is liberated in the form of energy—99·3 to 99·9 per cent. of their substance remaining in the form of matter. The mysterious cosmic rays that bombard the earth from outer space, do so with energy billions of times greater than that released by fission or fusion. If we ever discover how to create matter from energy or how to liberate anything like one hundred per cent. of the energy locked up in the atom, we shall have powers within our grasp that stagger the imagination. It rests with us how we use even the powers we already have. With them we can create a wonderful new world for ourselves far beyond anything hitherto imagined—*or we can destroy it.*

All life on the Earth is entirely dependent on the Sun for its well being and continued existence, even for the fact that it exists at all. Any increase or diminution of the radiation given out by the sun, even though it were comparatively slight, would mean that the life of the Earth would be wiped out. The very existence of life from its first inception, and its continuous evolution from earliest microscopic beginnings has been rendered possible and controlled by solar radiation. The progress and wellbeing of the human race ever since the earliest men evolved from ape-like ancestors, has been due to the Sun. Little wonder then that with the first early dawning of intelligence, men began to worship the Sun as a deity to whom they were indebted for everything, even life itself.

The question of how life originates is still unexplained. It seems probable that the development of carbon compounds over a long period leads up to the production of amino-acids and proteins. This and the action of mineral catalysts, would appear to be essential first stages in the development of living from non-living matter. We may assume that the earliest organisms existed in water without free oxygen, and since most of the free oxygen of our atmosphere is provided by the activities of green plants, it would seem that vegetable life must have preceded animal life.

All green plants and a few bacteria are autotrophic, which means briefly that they are not dependent on outside sources for their organic food supplies, but are able to manufacture their own organic material from inorganic ones, with the help of energy absorbed from their environment. In the case of the green plants, organic material is manufactured through the agency of the green pigment known as chlorophyll, from water and carbon dioxide, with sunlight providing the energy needed for the carrying out of the process. In the case of certain bacteria, the energy required for the synthesizing of organic materials is supplied by oxidation processes, of hydrogen sulphide by the sulphur bacteria, and of hydrogen by the hydrogen bacteria. They are what is known as chemo-autotrophic and chemo-synthetic.

The non-green plants, a few of the higher orders of plant life, the fungi, most bacteria and all the animals including human beings, are parasitic organisms entirely dependent, directly or indirectly, on the green plants for their food. They are what is called heterotrophic, that is to say, they obtain organic substances (food) from other organisms living or dead. It is clear, therefore, that if all the green plant life of the world were suddenly destroyed, by an epidemic, for example, the entire animal life of the Earth would quickly perish. The first to go would be the plant-eating mammals, then the carnivorous animals, which includes the human race, and, lastly, the carrion birds, crows, vultures, etc. If, on the other hand, the entire animal population of the world died out, the green plant life would continue indefinitely. It is clear from this that the autotrophic character of the green plants is the basis of all life, animal and vegetable. The plants are able to create, while the animals can only consume that which has been created for them.

The process by which the chlorophyll-bearing plants, as well as a few very primitive green animals, the fresh water protozoon, for example, are able to produce their own organic material is known as photosynthesis. This may be briefly defined as a reaction of carbon dioxide with water as a result of which carbohydrates are built up under the catalytic action of chlorophyll, solar energy being employed in the process. As a result, six molecules of water and six molecules of carbon dioxide are transformed by the aid of light energy from the Sun into one molecule of glucose (sugar), and six molecules of oxygen. The glucose so produced is organic material, while oxygen is liberated and, by a process of metabolism in the plant, proteins and fats are formed from the carbohydrates.

Since the green plants (photoautotrophs) are not dependent on other organisms for existence but merely on their physical environment, their hold on life is much surer and more stable than is that of the animals (heterotrophs). They will, therefore, make their appearance on a planet long before conditions have become suitable for the higher forms of life and will enjoy continued existence long after all the higher forms have disappeared.

CHAPTER II

The Substance of Life

Martian Conditions—Life on Mars

LIFE HAS developed or evolved on this particular planet in the way it has, simply because it was obliged to adapt itself to conditions existing here. Had the conditions been materially different, we should either not be here at all or we should have developed along very different lines.

To take only one example, we may examine the effect produced by the small amount of ozone present in the atmosphere. Ozone is merely a heavy variety of oxygen, having three atoms to the molecule instead of the two of ordinary oxygen. Most of the ozone in the atmosphere is at a height of between fifteen and thirty miles. The amount present is exceedingly small but the absorption it exercises in the ultra-violet region of the spectrum is so powerful that light of any shorter wave length than 0·000012 inch is absorbed and thus fails to penetrate the ozone layer in the atmosphere. Were it able to do so, living things, including ourselves, would not be able to exist as now constituted nor could our eyes have developed in the way they have. The atmospheric ozone allows just enough ultra-violet light to pass as will be beneficial, while preventing the passage of an amount that might be injurious. If there was no ozone in the atmosphere, or much more than there is, life, as at present constituted, could not continue.

THERE IS LIFE ON MARS

This is but one example of the way in which life adapts itself to existing conditions. Had the conditions been different, life would have developed along different lines or it might, perhaps, never have come into existence at all. This is a point to consider when examining the possibility of life on other planets, having regard to conditions existing there, so far as we are able to determine them.

As we have seen, all matter is composed of the same basic elements associated in various ways, but we find that the substance of plants and animals (living substance) is built up of the lighter ones. The chief elements of Protoplasm (the life substance) are Hydrogen, Carbon, Nitrogen, Oxygen, Phosphorus and Sulphur. These are found in the form of proteins, fats and carbohydrates. Other elements which occur in small amounts are Sodium, Magnesium, Silicon, Chlorine, Potassium, Calcium, Iron and Iodine, occurring in the form of acids and salts. Carbon, as already mentioned, is the basic element, while Iodine is the only one of considerable atomic weight. The basic chemical of hæmoglobin in the red blood corpuscles of animals is iron. Without hæmoglobin, which enables oxygen to be absorbed into the cells, respiration would be impossible. Magnesium occurs in chlorophyll, the green colouring matter of plants essential for the process of photosynthesis by which the green plants are able to convert inorganic (non-living) substances into food.

Anabolism, as it is called, is the process within the life material (protoplasm) whereby substances possessed of little energy are converted into highly complex ones having a great amount of energy in a form that makes it readily usable. The reverse of this building up process, known as catabolism, is the process by which the complex substances built up are again broken down into simple substances; as a result of which process energy is released to be used in the bodily functions. In other words, it is the conversion of chemical energy into kinetic energy, movement, heat and other activities of the animal body. When the energy has been extracted from them the waste materials are discarded in the form of water, carbon-dioxide, urea, etc.

This process of building up and breaking down, known as metabolism, is simply a *means of transforming the energy of the Sun into the physical energy of living creatures.*

There are two methods by which catabolism may be accomplished: by fermentation or by combustion. The first is the means by which energy is liberated in primitive organisms such as yeast. It is the least efficient method since the break-down of the molecule into its basic elements is incomplete and, therefore, the gain of energy comparatively small. In combustion, oxygen is the agent employed. This is the most efficient form of catabolism we know as it releases the greatest amount of energy from the matter consumed.

It is not impossible that other more efficient means of converting solar energy (the energy of a sun or star) into the energy displayed by living things may exist elsewhere in the Universe. If so, there may be forms of life entirely beyond our understanding or imagination.

Respiration is the principal means by which organisms are able to achieve the chemical reactions for building up and consuming energy. Both plants and animals need oxygen for the process of combustion necessary to release energy.

In the case of very primitive and minute organisms,

the body surface affords sufficient area for breathing, as they do, through the skin. Larger and more complex creatures require gas in quantities impossible to obtain in this way. A greater breathing surface is, therefore, obtained by the folding of tissue in certain areas. In fish, these localised areas take the form of gills which are merely outer extensions of the inner tissue. A much larger area for respiration is obtained in the case of most land animals, birds and insects, by an expansion of the surface fold, inside the body: lungs in the case of the vertebrates and a tracheal system in insects.

The trachea is a system of tubes which start at the skin and branch into smaller tubes, ending in, or between, the body cells.

Lungs are found in three different stages of development, starting with a simple sac in the case of newts; a sac divided into separate compartments in frogs and, lastly, in warm-blooded creatures a complicated structure of air cells in bladder-like form. These present a very large surface to the air entering the lungs. Both lungs and tracheæ make it possible for the animal or insect to maintain air inside its body, air which is materially different in composition from that outside the body; a kind of internal atmosphere as it were. Plants are able to maintain a similar internal atmosphere within their leaves. This principle is essential to plants or animals, enabling them to adapt themselves to changing conditions or a different environment.

Life is a form of combustion. The combustion that consumes the food materials in the animal body is to all intents and purposes the same process as occurs in a bonfire, or any other fire, except that, in the case of the animal, the combustion takes place at a much slower rate. If it were not so, the materials would be consumed too rapidly for the organism to convert them into physical activity. This fact is of considerable importance when it comes to assessing the chances of life, as we know it, occurring on other planets where climatic and atmospheric conditions differ widely from those prevailing here. On Mars, for example, owing to the small amount of oxygen present in the atmosphere, it would be impossible to light a fire or even to strike a match, for without oxygen in sufficient quantity, combustion cannot take place.

PLANETARY ECOLOGY

An ecological space may be defined as one in which living things can exist. Such a space, even though it may be free from factors that would tend to destroy life, will not provide a suitable home for living organisms if the conditions favourable to life are present either in too great or too small a quantity. The principal factors applicable to and essential for the existence of life in any of the forms with which we are familiar are temperature, light, humidity and the chemical constituents of air, soil and water. The most important factor governing the suitability of an environment for life is probably temperature.

The range of tolerance towards their environment varies enormously with different organisms. Certain tropical plants, for example, can withstand a temperature range of no more than 20° F., while some of the lower forms of plant life, such as lichens, can tolerate a range of as much as 180° F. Certain bacteria found in hot springs in America and elsewhere carry on active life at a temperature of 170° F., while trees in Northern Siberia and some Arctic plants are able to survive at a temperature of 75° F. below zero. Some of the lower organisms, bacteria, algæ, lichens and mosses, can survive for lengthy periods at temperatures approaching absolute zero. Above or below temperature that permits of active life, the organism passes into a dormant state, from which it will emerge with the coming of improved conditions, *e.g.*, a rise or fall of temperature.

The scale in general use for measuring absolute temperatures in physics and astronomy is that known as the Kelvin. Starting from absolute zero at 460° below zero F. it has a top range of almost a million million degrees F. On this scale the range over which active life can exist is little over 110°. Below this range the movement of the molecules slows down until all movement ceases at absolute zero. Above it the molecules break up. At more than 5,000° F. the shells of the atoms start to crack, and above 20 to 35 million degrees F., even the nuclei of the atoms are disturbed.

The normal temperatures on Earth range from approximately 75° below zero F. to about 140° above, while on Mars the range is from 95° below zero F. to about 87° above. The only other planet in the solar system with a temperature scale covering the range of active life is Venus. Venus moves in an orbit at a mean distance from the Sun of 67 million miles, the Earth 93 million miles and Mars 142 million miles. Since the intensity of radiation decreases with the inverse square of its distance from the source, it is obvious that Venus must be a very much warmer planet than the Earth, and Mars a very much colder one. The intensity of radiation received by Venus is nearly twice that received by the Earth, while Mars receives less

THE SUBSTANCE OF LIFE

than half as much. The lowest temperature ever recorded on the Earth was 90° F. below zero, in Siberia in 1892, and the highest 136° F. near Tripoli in 1922.

The Martian climate, although so much colder than ours, is more equable; that is to say, it is not subject to sudden changes of temperature or storms such as we experience here. The reason for this is that there are no seas and no high mountains. If the world were all sea, or one great plain with no high mountains and no seas, there would be no storms, merely gentle winds up to perhaps twenty-five miles an hour, such as occur on Mars. On the other hand, living things on Mars have to withstand temperatures ranging up to a maximum of about 85° F. by day and falling to a minimum as low as perhaps 30° to 90° below zero F. at night.

Naturally, the temperature range between the Martian day and night is not always so great since it varies with the time of year and the distance of the planet from the Sun. Radiometric temperature measurements have shown a variation in the southern tropical region of between 87° F. by day and − 40° F. by night.

A factor closely connected with that of temperature is the amount of oxygen present in the atmosphere, for on this depends the sort of life we may expect to find on a planet, always assuming that we are not dealing with some type of life entirely outside our experience, and following principles which are beyond our present knowledge or understanding.

Here on Earth we are accustomed to classify living things in two categories: (1) the cold-blooded, which comprises plants and creatures up to and including reptiles, and (2) the warm-blooded, birds and mammals, including the human mammal.

The first named are dependent on the outside atmosphere for the maintenance of their body temperature, becoming active with warmth and torpid with cold. The temperature of the birds and mammals is regulated by an internal physical and chemical mechanism requiring a high metabolism to generate heat. They, therefore, require much more oxygen than do cold-blooded creatures.

Now oxygen is essential to life as we know it, being the chief agent for biological oxidation, which releases energy, and on Mars free oxygen is either not present at all or, if present, available only in minute quantity. This oxygen scarcity would rule out the possibility of any of the higher forms of life, *as we know them*, existing there.

Atmospheric pressure is measured with a mercury barometer calibrated in millimetres. The total atmospheric pressure at sea level is 760 mm. Hg, Hg being the symbol of mercury. The pressure falls with increase of altitude until at about 30 miles up it is less than 1 mm. Hg.

The atmospheric pressure at the surface of Mars may be approximately 60 to 70 mm. Hg, roughly the equivalent of that prevailing eleven miles above the surface of the Earth. The oxygen content of our air is about twenty-one per cent. of the total pressure, therefore the oxygen pressure at sea level is 160 mm. Hg. The proportion is fairly constant up to an altitude of about 55 miles so that the oxygen pressure at any level can be calculated from the total atmospheric pressure at the same height.

The oxygen pressure at the surface of Mars is probably about 1 mm. Hg, and to obtain the equivalent on Earth we should have to rise to a height of about a hundred thousand feet. The minimum oxygen pressure

THE SUBSTANCE OF LIFE

that human beings can endure is 65 mm. Hg, at which stage they begin to lose consciousness. At 10 mm. Hg a candle flame will not burn, and no fire could be lighted.

Plants breathe and consume oxygen and some of them will continue to breathe until the oxygen pressure falls to below 1·5 mm. Hg. They have the capacity, however, to adopt anærobic* respiration at any time. In order to thrive and develop, plants require a considerably higher oxygen pressure than exists at the surface of Mars, but they have a way of overcoming this lack of pressure, possessed by no other living things. As we have seen, green plants, so long as they are exposed to sunlight, are able to produce their own oxygen by photosynthesis. It has been estimated that the green plant life of the world produces about five thousand million tons of oxygen in a year.

The lowest temperature at which photosynthesis can be carried on by the majority of plants is 32° F., the freezing point of water. Some lichens, however, growing in the Arctic are able to carry on photosynthesis at temperatures approaching zero Fahrenheit. The temperature on Mars and the amount of sunlight available are amply sufficient for the process. Carbon dioxide is more plentiful on Mars than on the Earth and water is available although, perhaps, mainly in the form of snow and vapour and both water and carbon dioxide are essential for photosynthesis.

The fact that water may be available mainly as vapour would be no obstacle to certain types of vegetation. Some terrestrial lichens are very hygroscopic. In other words, able to absorb moisture readily from the atmosphere.

* Anærobic—living without air.

THERE IS LIFE ON MARS

The plant life of Mars may, therefore, be confined to the lower orders of cellular plants, lichens, mosses and algæ, unless, and it is not impossible, the Martian plants follow principles that are as yet unknown to us.

Lichens are interesting for several reasons. For one thing they are the hardiest plants known. They are dual organisms formed from the symbiotic association of two plants, a fungus and an alga. Symbiosis is the association of dissimilar organisms to their mutual advantage; for example, the association of nitrogen-fixing bacteria with leguminous plants, peas, beans, etc. The bacteria, which inhabit nodules on the roots, manufacture nitrogen compounds from the nitrogen of the air, making these available to the plant. In return the plant supplies them with carbohydrates and other food materials.

In lichens, the fungus partner offers protection from cold and provides inorganic substances, among them water, due to the extremely hygroscopic character of most fungi. The algæ partners, for their part, build up organic substances and also supply oxygen by means of photosynthesis. As a consequence of this association, lichens are very resistant to cold and can survive in very dry conditions. They grow readily on tree trunks, walls, barren lava and exposed rock, as well as on the ground and supply the dominant vegetation in high mountain and arctic regions. The organic acids which they produce decompose rock and stone, so making humus available for less hardy plants. Together with liverworts and mosses, they play an important part as pioneer vegetation in barren areas. They may well have been among the first vegetation to appear on the Earth spreading over the barren volcanic rocks,

THE SUBSTANCE OF LIFE

and they might be the last living things on a planet from which all other life has departed.

The vegetation which appears and is observed to spread out on the surface of Mars in the spring, does so with great rapidity, from which we must assume that it has an intensive metabolism, requiring oxygen for respiration, and oxygen is anything but plentiful in the Martian atmosphere. The question thus arises as to how the vegetation obtains oxygen in the required quantities. The answer lies in the mechanism that enables plants (and animals) to maintain an internal atmosphere.

The system of intercellular air spaces within the tissue of a plant enable it to maintain an internal atmosphere that makes existence easier in an atmosphere that is poor in oxygen. Terrestrial plants are able to breathe at night, when photosynthesis is suspended, because there is plenty of free oxygen in the atmosphere. On Mars, where there is little oxygen, and the nights are intensely cold, it may be that the plants store up most of the oxygen they have produced through photosynthesis during the day, for use as required, instead of adding to the free oxygen in the air as do terrestrial plants. The night temperature probably induces a dormant condition in the plants which, since they would thus be expending no energy, would require no oxygen. The Martian plant life could, in this way, survive through being active by day and dormant by night. There is nothing very strange about this for our own vegetation is dormant in winter, and each night on Mars may be regarded as a brief winter. As practically all our terrestrial supply of oxygen is due to the activity of plants, it may well be that the early life of the Earth was intermittent, *e.g.*, active by day and latent by night. *Free oxygen may not be a necessity for the genesis or development of life in its earliest stages.*

Since it can produce its own oxygen supply, the really essential gas required by vegetation is carbon dioxide, and this, we know, exists in greater quantity on Mars than on the Earth. There is also the possibility that a vast store of oxygen is locked up in the soil of the planet, which some authorities believe to consist largely of ferrous oxide of limonite. The colour of much of the surface of Mars leads to this conclusion and it also points to the probability that free oxygen must have been plentiful at some time in the past.

It may be that the vegetation, existing there, is the last remaining vestige of Martian life that still clings to portions of the surface and manages to survive. Certain types of bacteria probably exist. Some terrestrial bacteria are independent of sunlight and derive their energy from inorganic matter such as hydrogen, ammonia, methane, etc., by oxidation. There may be enough oxygen in the atmosphere for creatures of this primitive type, or they may obtain oxygen and nourishment from the vegetation. Bacteria of this type possibly existed in the proto-atmospheres which enveloped the planets, including the Earth, in their early stages and they could have been among the first living things in the world.

CHAPTER III

What is Life?

TERRESTRIAL LIFE began in water and probably remained there for hundreds of millions of years before it was able to invade the land. The early organisms must have existed in water without free oxygen and it was not until very primitive forms of plant life evolved, that an atmosphere in which animal life could thrive was brought about. Certain one-celled plants are independent of oxygen although they, themselves, emit it, and it may be that the earliest plants were something of this kind which, by producing oxygen, rendered the atmosphere breathable for animals. Nearly all the free oxygen that now exists has been produced by the green plant life of the world.

Amino acids are organic compounds which contain basic amino (NH_2) and acidic carboxyl (COOH) groups. Hundreds or thousands of amino acid molecules go to build up each protein molecule. They are thus fundamental constituents of living matter. Some twenty-five different amino acids are known as constituents of protein.

Proteins are very complex organic compounds built up of numerous amino acids, present in all living things. Their basic importance is clear from the fact that both enzymes and genes are proteins. As already mentioned, a protein molecule may be made up of hundreds or thousands of amino acid molecules. The possible arrangement of amino acids appear to be almost limitless, and every living species, possibly every individual, may have protein molecules of a kind peculiar to itself.

Every molecule of protein is made by another protein molecule, and other living molecules are built up on protein. The question is, therefore, how did the first proteins arise? To this we can only hazard the theory that they may have been built up on wet mud or clay, so beginning the chain of life. They can no longer arise in this way; firstly, because any such living form newly produced would be immediately eaten by living organisms already existing and, secondly, because energy that built up such molecules in the past, the short wave radiation from the Sun, is now prevented from reaching the surface of the Earth by the layers of ozone brought into existence by terrestrial vegetation.

Without green vegetation animal life is impossible and, as we know, without bacteria no life at all would be possible on the Earth. About ninety-five per cent. of the carbon dioxide needed by the higher plants on which all the higher forms of animal life are in turn dependent, comes from microbial metabolism.

The discovery of *anærobic bacteria*, a form of life that can exist without air, throws some light on the probable earliest forms of terrestrial life, and must, to some degree, modify our views on the extent to which life may be capable of adapting itself to conditions existing on other planets having atmospheres very different from ours.

Photosynthesis, the process of building up complex organic chemical substances from carbon dioxide and water, can only operate through the influence of light

WHAT IS LIFE?

obtained indirectly from sunlight. In the process of photosynthesis oxygen is released. While photosynthesis is taking place, the plant is breathing by taking in oxygen and giving out carbon dioxide, the reverse process. During daylight the process of photosynthesis proceeds at a very much faster rate than breathing, the result being that great quantities of oxygen are produced by the plants during the day. Green plants are the only living things that can build up food substances from the carbon dioxide which we, and all other animals, breathe out, and water. These complex chemical substances are then made available for grazing animals and also for fungi and bacteria, none of which can build up food substances for themselves as can the green plants. The grazing animals are eaten by carnivores which, in turn, may, themselves, be eaten by other animals. The flesh, even of carnivorous animals, is thus built up indirectly from grass.

All our food, therefore, comes from green plants or from animals that have eaten green plants, or from animals that have eaten animals that have eaten green plants. In other words, *life is parasitic upon itself*.

It is *chlorophyll*, the green pigment found in all types of plants except fungi and a few flowering plants, that enables green plants to build up carbohydrates from carbon dioxide and water, while absorbing energy from sunlight in the process of photosynthesis. We may regard the autotrophic character of green plants as the basis of life.

The life substance—*protoplasm*, is a jelly-like, complex compound, made up of ordinary chemical elements, principally carbon, nitrogen, hydrogen and oxygen with traces of sulphur, phosphorus, iron, calcium, potassium, sodium and chlorine. Protoplasm is the living substance within all the cells of every living thing, including the plasma-membrane of the cells. Without it we should not be alive. If it is destroyed, we are dead. To kill us it is not necessary to destroy *all* the protoplasm in *all* the cells. The destruction of a few vital cells is enough, the rest will die too, although not necessarily all at once. Portions of us may go on living after what we call death has occurred.

The genesis of life and its continuous evolution may be merely the logical continuation of a vast cosmic process in the course of which energy builds up into atoms of hydrogen from which, in turn, everything else in the Universe is built up. Life is basically an electro-chemical phenomonen. It may well be, therefore, that the basic force behind all the manifold workings and wonders of the Universe is electrical energy, at once eternal and indestructible. The atom, from which all things from the stars to the most minute specks of living matter are built, is electrical.

It is now known that microbial protoplasm is some twenty times as plentiful on Earth as that of the higher forms of life. This is due to the omnipresent character of bacteria. The study of microbial life has been of great value through the insight it has afforded us of the flexibility of the life process in the living cell, and its ability to cope with a changing environment. It has also thrown much light on the underlying causes of evolution.

BIO-CHEMISTRY

Bio-chemistry may be defined as the science that concerns itself with the study of the living process. Physics and chemistry deal with the reactions between matter

WHAT IS LIFE?

and matter, while bio-chemistry is, as it were, a special study of chemistry.

Up to a few years ago it would probably have been denied that the life process could be reduced to terms of chemistry, and the re-actions of living matter studied scientifically.

So long as its heart beats, blood circulates and it breathes, an animal is alive. So is a plant while its leaves are fresh and green and its flowers open to the warmth and light of the sun. Shoot the animal through the brain or heart and it lies motionless, rapidly cooling and stiffening. Uproot the plant and it quickly shrivels and dies.

If, however, we cut a stem from the uprooted plant and place it in water, the leaves will remain fresh and the flowers will continue to open for some days; or if we cut a piece out of the animal while it is still warm and keep it in a special mixture of mineral substances and foodstuffs dissolved in water maintained at blood heat, it will continue growing as if it were still a part of the live animal. The flower or the animal tissue go on living although the plant or the body from which they were cut is dead. This is simply because we have given them conditions that resemble, to some extent, those that existed before they were severed. The hearts of animals have been removed from the body and kept alive for years by giving them conditions approximating to those existing while they were still contained in the animal body, *e.g.*, pumping blood through them at the correct temperature, etc. It makes no difference whether the heart is that of a chicken, a dog or a man.

It is clear, therefore, that life is not just a matter of a beating heart, breathing lungs, the presence of a soul or spirit, or of roots fast in well watered soil.

THERE IS LIFE ON MARS

At a very early stage in the evolution of life, the only living things were single-celled microscopic organisms living in water, and it was not until such organisms in some way developed the ability to coalesce into groups or masses of living material that anything that was not microscopic or ultra-microscopic was able to evolve.

The human body contains something like a million million cells, and they have all come by division from a single original cell. Every living thing, whether it is a man, an oak tree, a cabbage or a cod fish, starts life as one microscopic cell. In a single-celled organism such as the amœba, all the functions: absorption of food, evacuation of waste matter, movement, response to stimulus and so on, have to be performed by the single cell, and this is the case with each cell of some primitive multi-celled organisms. In the case of the higher multi-celled animals or plants, the work of the cells is divided up, each cell performing certain functions. This division of work in multi-celled animals or plants results in the development of specialised groups of cells, which may be compact as in muscles, or diffuse as in blood cells (corpuscles). Each of these groups of cells is known as a tissue, and one or more tissues are so arranged as to form organs. In this way, nerve, muscle, connective tissue and skin are combined to form the various organs of the body, heart, lungs, liver, kidneys, and so on.

Bio-chemistry concentrates on examining life scientifically, as a natural phenomenon displayed by matter, so constituted, that it reacts with other matter in such a way as will enable it to grow and reproduce itself. The work of the biologists has confirmed the reality of the theory of evolution and the unity of all forms of life,

WHAT IS LIFE?

animal and vegetable, while that of the bio-chemists has lent added confirmation if any were needed.

The cells of each particular tissue have a similar function and a similar structure. These differ, quite obviously, from both the function and structure of cells of other tissues.

From the very first the bio-chemist has to avoid any tendency to lay too much emphasis on the immense variety or diversity of living things and to concentrate more on the similarity of all living tissues. The proteins derived from parts of the human body are constructed from the same amino acids as are the contents of a bacterial cell. Certain bacteria require the same vitamins for their growth as do certain mammals, although to outward appearances they are such entirely different organisms. The similarity of structure and material between living entities results from the conditions imposed upon them by the planetary conditions to which they are all subject and which tend to produce uniformity. The seeming differences that do exist, such as, for example, the basis of the skeleton, bone in the case of the vertebrates, cellulose in that of plants and chitin in that of insects, may be more apparent than real, and due more to the inadequacy of the knowledge so far gained than to any fundamental differences. The discoveries of the bio-chemist all tend to confirm that life is more remarkable for its similarities than for its differences and that, in whatever form it may exist, it can only be regarded as a single uniform phenomenon.

We are only, as it were, at the very beginning of the acquisition of knowledge with regard to life. Even the reason why death from old age should occur in the higher forms of life is not clear. We know that the tissues cease to operate but there is no obvious reason why they should do so, nor why they should not continue to function or retain their youthful vigour. Some of the plant viruses such as those which attack the tobacco plant or the potato, are among what are known as *filter-passing virus* since they will pass through filters that retain bacteria. Similar organisms cause certain diseases in animals. That they are living things is shown by the way in which they multiply in plants or animals, and that heat destroys the infectivity of the filtered material. Nucleo-proteins, substances allied to proteins, have been isolated from the leaves of infected plants, and these have the infective properties of viruses. If a plant is infected with small amounts of these, larger amounts may be recovered, proving that the substances have the property of reproducing themselves when introduced into a suitable medium. Away from the plant cells they show no apparent signs of life. Clearly these substances are very near the border line separating living from non-living matter, and their existence lends added weight to the theory that life is a special property of matter arranged in a particular way. They cannot, however, be regarded as a first or even an early step in the evolution of life since they are dependent for their existence on other highly organised forms of life. Single-celled plants may be regarded as an earlier product of evolution.

It is easy to decry or belittle the efforts of the scientists to explain the nature and origin of life. Immense progress has and is being made towards this, although it may be a very long time yet before we have a definite answer to the problem. Enough, however, has been accomplished to show that it is not an unsolvable one, or that anything that might be termed super-normal, or super-natural is involved. Generations of scientists may yet have to come and go before the question of

WHAT IS LIFE?

the origin of life is finally solved. That it will be solved eventually is as certain as anything can ever be amid the uncertainties that surround us. No one can say with any certainty that the Sun will not explode one day and destroy us with all our works. Other stars are observed to do so from time to time. So far as we know it is most unlikely to happen but we cannot say it is a complete impossibility.

Micro organisms are omnipresent, that is to say they occur practically everywhere in the world. Their rate of multiplication is usually enormous and their numbers infinite. Their importance is immense. It is no exaggeration to say that without them life on Earth would be impossible. If all the bacteria suddenly disappeared there would be an end to many diseases that now afflict plant and animal life. At the same time dead animals and vegetables would not decay but would lie about cumbering up the earth. Leguminous crops which now restore nitrogen to the soil would become useless for this purpose, and ruminant animals would die of starvation as a consequence, so that very soon all life would disappear.

The natural fate of plants and animals when they die is to lie on the surface until they disintegrate so that the materials from which they were originally built up can be restored to the soil and atmosphere. Even the skeleton soon disappears as a result of bacterial activity. That is why skeletons, contrary to popular belief, are seldom found by divers in wrecked ships. Burying the body in a coffin six feet deep in the earth merely delays the process of restoring its chemical constituents to earth and atmosphere, while cremation, if the ashes are preserved in an urn, robs nature of part of its due, until such time as they are scattered to the winds.

CHAPTER IV

Factors in Planetary Temperature Variations
The Terrestrial Glaciations

THE TEMPERATURE at the surface of planets is not only dependent on such circumstances as their distance from the Sun and the eccentricity of their orbits, but also on the amount of heat given out by the Sun, which may vary appreciably over a period, and the composition of the planetary atmospheres.*

The temperature of the Earth has varied considerably from time to time; the Glaciations or Ice Ages, with the intervening warm periods, are proof of this, but the reasons behind these great changes in temperature are not, as yet, clearly understood.

For one thing, the amount of carbon dioxide present in an atmosphere can make a very appreciable difference to the temperature of a planet, and we know that both Mars and Venus have considerably more carbon dioxide in their atmospheres than has the Earth.

While variations from time to time in the solar temperature, and the amount of radiation given out by the Sun as a consequence of such variations, may be the cause of the recurrent glaciations and intervening warm periods on the Earth, there are other phenomena that might be responsible or at least be contributory causes.

---

* It will be appreciated that certain of the questions touched upon in this chapter are at present speculative rather than factual. Some of the possible causes of temperature variation on the Earth are given purely as suggestions.

PLANETARY TEMPERATURE VARIATIONS

The Earth now appears to be in the early stages of one of the periodic warming-up eras, and statistics prove that, even during the last hundred years, the average temperature of the Earth has risen by about 2° F. Furthermore, most of this increase has occurred during the last fifty years or so. This rise of 2° F. may strike most people as insignificant, but it should be borne in mind that a fall of only 4° F. in the average temperature of the Earth would suffice to bring back glacial conditions over large portions of the world, so that eventually much of Europe, Asia and North America would once again be covered by vast ice sheets advancing from the North Polar regions. If, on the other hand, as seems likely, this same rate of warming-up continues, it will lead to a raising of the sea level caused by the melting of the Arctic ice fields. Over so short a period as the next fifty years this could result in considerable flooding of certain low lying coastal areas, and over a much longer time, the submergence of much of what is now dry land. Given a long enough time it could change the whole geographical face of the Earth.

The age of the Earth is believed to be some three thousand to three thousand five hundred million years, and during that period the temperature has been rising and falling continually, although during most of its existence the Earth has been very much hotter than it is now. Coal is found not only in England and North America, but also in Greenland, Siberia, Alaska and the Arctic regions, and coal is merely the fossilised remains of tropical vegetation that once covered the surface. The glaciations or periods of intense cold have been relatively short, a few million years each, with much longer warm periods intervening.

In some parts of the world, where records have been

kept, the rise in temperature over the last century has been as much as 4° F. In Greenland, Alaska and other areas, long warm summers and short, mild winters are causing a recession of the ice sheets that once covered the whole of these parts, and even extended as far south as Germany, France and much of North America, little more than 20,000 years ago. As we have seen, these changes may be due to a heating up of the Sun, but they may equally be caused by a tilting of the Earth's axis, *an increase or decrease of the carbon dioxide in the atmosphere*, or to more or less activity of the Earth's volcanoes.

Volcanic activity, while it is not a very likely explanation, may, nevertheless, be at least a contributory cause of temperature variation. A really big eruption such as that of Krakatoa in 1883, hurls immense quantities of very fine ash into the atmosphere. This is carried by the winds to all parts of the Earth with a consequent reduction in the amount of sunlight reaching the surface. In some places a fall in the average amount of sunshine amounting to as much as 12° F. was recorded following the Krakatoa eruption. This was the last really big eruption of modern times and consequently the air may be clearer now and the amount of sunshine reaching the Earth's surface correspondingly increased.

The theory that an increase or decrease in the amount of carbon dioxide in the atmosphere is responsible for changes in the Earth's temperature was advanced by a British scientist, John Tyndall, nearly a hundred years ago, and it has recently been further developed by Professor Gilbert Plass of America. It is based on the fact that the air normally contains 0·03 per cent. of carbon dioxide, which is transparent to the direct rays

PLANETARY TEMPERATURE VARIATIONS

of the sun and which absorbs them. In this way they are trapped instead of escaping into outer space, so that they remain to warm the atmosphere, and consequently also the Earth's surface. An increase in the amount of carbon dioxide present in the atmosphere would, therefore, lead to an increase in temperature. Scientific measurements of the carbon dioxide content of the atmosphere show that during the last fifty years it has increased by some ten per cent. This would account for the amount by which the Earth's average temperature has increased during the same period.

There are several possible reasons why the amount of carbon dioxide present in the atmosphere should vary from time to time. We have already examined the process by which green plants absorb carbon dioxide from the air, and with this and water, under the influence of sunlight, manufacture sugars, starches and cellulose and so build up their structures. In the course of this process something like a million million tons of carbon dioxide are taken from the atmosphere annually. It is returned to the atmosphere, however, when the plants and the animals which feed on them die and decay and is only lost when, as in the case of coal, vegetation becomes buried in the ground. The loss from this and other causes, however, is so small as to be comparatively negligible.

Another process by which carbon dioxide is removed from the atmosphere is the constant weathering of the rocks and their conversion into soil. This change is, to a great extent, effected by carbonic acid which forms when carbon dioxide is dissolved in water. The acid attacks the rocks bringing about their conversion from silicates to carbonates, a slow process that removes about a hundred million tons of carbon dioxide from the atmosphere annually. This loss, however, is just about balanced by the considerable quantities of carbon dioxide released into the air every year by volcanoes. The two processes are not constant, however, and a rise or fall of the Earth's temperature may come about from the rate at which carbon dioxide is removed or added to the atmosphere from time to time.

Yet another factor that may be responsible for a gradually increasing temperature of the Earth, is the amount of carbon dioxide released into the atmosphere by the burning of coal and oil, which, it is estimated, adds each year about six thousand million tons of carbon dioxide to the air. A change in the carbon dioxide content of the atmosphere, besides affecting temperature, also affects humidity. A cloud gives up its moisture and rain falls when there is an appreciable difference in temperature between the upper and lower surfaces. When there is less carbon dioxide present in the atmosphere, the upper surface of the cloud tends to cool more quickly and the rainfall is thus increased.* Cooler and wetter periods occur when the carbon dioxide content of the atmosphere is low, while hot dry periods are experienced when the carbon dioxide content is high. It looks as if we are living in a period of gradually increasing temperature and dryness, but how long this process will continue no one can say. Over a long period it could have far-reaching effects. Canada and Siberia might eventually have temperate climates, while the United States, Britain and most of Europe might become sub-tropical.

* This question of carbon dioxide is of great importance, as we shall see presently when we come to examine the probable composition of the atmosphere of Mars and other planets.

PLANETARY TEMPERATURE VARIATIONS

In some places the rise in temperature is sensational. Spitzbergen, for example, shows a rise in the average winter temperature of 18 degrees F. over the last fifty years, and the harbour there, which was formerly ice bound for the greater part of the year, is now open for nearly seven months annually. The northern ports of Russia and Siberia show a similar tendency and are free of ice for a much longer period than has ever been known in the past. The glaciers are receding and getting smaller from Siberia to Switzerland and from Iceland to Greenland. Large areas of Greenland, which were formerly covered by ice, are now growing crops and the trees and vegetation are moving northwards at an ever growing pace. Animal life, too, is migrating north as a result of the increase in temperature and the disappearance of the permanent covering of snow and ice. The seas, too, are getting warmer with the result that the fish are also migrating farther north. The British trawler owners now find that their boats have to go much farther afield than they did previously if they are to obtain profitable catches.

When vegetation obtains a hold on ground on which nothing has grown for perhaps thousands of years, it proves extremely tenacious and can withstand severe annual extremes of temperature. Moreover, the vegetation in itself tends materially to increase the temperature, partly due to the fact that, while the heat of the sun is reflected away by a covering of snow, the darker vegetation absorbs the warmth of the sun and allows it to soak into the ground. The increase of vegetation also helps to increase the warming-up process of the sea, for as the sea warms up, so the primitive algæ, which are seen as a green scum in ponds and ditches, increases more rapidly and this increase, in turn, helps the seas and oceans to store up more heat and so the warming-up process gains added impetus.

As we have seen, the burning of coal and oil increases the amount of carbon dioxide in the atmosphere so that present day activities of man may be responsible for a material increase in the warming-up process. The main danger, if this warming-up continues for any considerable length of time, will be the flooding of large areas of what is now dry land due to the melting of the Arctic ice. It has been estimated that if all the ice at the North and South Poles melted, it would result in a rise of sea and ocean levels of something like 200 to 250 feet. This would bring about a major catastrophe and would change the whole face of the Earth. Scientists consider, however, that any considerable recession of the ice in the Antarctic regions is unlikely owing to the intense cold prevailing there, a cold far greater than that known at the North Pole. Another danger that may come from a rising temperature is increased evaporation from the ground leading, eventually, to a severe shortage of water.

Since we do not know the actual reasons for the warming up of the Northern Hemisphere, it is impossible to say whether it is likely to continue indefinitely. If it does, it will eventually have serious consequences for large areas of the Earth.

CHAPTER V

The Velocity of Escape
Planetary Atmospheres

THE VELOCITY of escape has an important bearing on the kind of atmosphere a planet may be expected to possess. The Earth's atmosphere is made up of countless billions of molecules of oxygen, nitrogen and other gases, all of which are darting about at enormous speed continually colliding with each other and rebounding in various directions. The lighter the molecule the greater its speed. The average speed of a molecule of oxygen is about a quarter of a mile per second, that of a hydrogen molecule about one mile a second. The velocity of molecules is dependent on the temperature of the gas. At absolute zero ($-273°$ C.) the molecules will be at rest having no velocity at all. The number of molecules in a cubic inch of air at normal temperature and pressure is something in the region of five hundred million billion, and under these conditions the average distance a molecule travels before colliding with another molecule is about the two hundred thousandth part of an inch. The distance travelled between collisions increases with reduced pressure of the gas, and since the density of the gaseous atmosphere of a planet falls with increasing distance from the surface, it is clear that in the upper regions of the atmosphere molecules will travel greater distances between collisions. In the outer regions of the atmosphere should a molecule rebound in an outward direction following a collision, if it does not collide with another molecule, it is possible that it may escape into outer space and consequently be permanently lost to the atmosphere. In order to do this, it must, of course, attain a speed greater than the velocity of escape.

The velocity of escape for the Earth is 7·1 miles per second, that of Mars 3·2, of Venus 6·5, of the Moon 1·5. By comparison, the figure for Jupiter is 38 miles per second, Saturn 23 and Uranus 14, while that for the Sun is 392.

Clearly, then, the type of atmosphere a planet may be expected to have, and the extensiveness, must be largely dependent on its size and consequently its velocity of escape.

The present atmosphere of the Earth is not the original one it started with. The primeval atmosphere must have been lost at a time when the Earth was very hot, indeed more than one primitive atmosphere may have been lost before the surface cooled sufficiently for the present one to become established. While the molten earth was gradually cooling down, enormous quantities of carbon dioxide, water vapour, and other gases would have been thrown up, and still more must have been added by volcanic action in the course of ages. These, together with some other gases still remaining from what was left of the initial atmosphere, formed the new atmosphere which the Earth, having cooled sufficiently, was able to retain. This atmosphere differed considerably from the present one in that it contained vast quantities of carbon dioxide and water-vapour but very little oxygen. In the course of time, as the Earth cooled, the water-vapour condensed out of the atmosphere and formed the oceans. Now oxygen, unlike the other gases in the atmosphere which are inert and do not form compounds with other elements, is chemically active and readily combines with other elements to form oxides; rust, for example, is simply iron oxide, while combustion is merely a process of oxidation which can only take place in the presence of oxygen.

As a result of this tendency of oxygen to combine with other elements, the oxygen content of the atmosphere is constantly being depleted. The most important example of this process is the weathering of the igneous or basic rocks which brings about the formation of sedimentary deposits; the weathered materials being carried away by rivers and streams and ultimately becoming the clay, sand or mud of the sea bed. As the weathering proceeds, fresh surfaces of rock, being continually exposed to the atmosphere, much of the ferrous oxide in the exposed surfaces becomes oxidised into ferric oxide, that is the red oxide of iron that we usually speak of as rust. It has been estimated that in the course of the geological ages, the thickness of the weathered deposits has amounted to at least four thousand feet and that the quantity of oxygen extracted from the atmosphere as a result would be nearly twice that now existing. The replenishment of the atmospheric oxygen, as we have seen, has come about through the agency of the green vegetable life of the Earth which absorbs carbon dioxide and gives out oxygen. At the same time the carbon dioxide supply is replenished and made available for building up plant cells through the decay of vegetable matter and other organic materials. The present abundance of oxygen in the atmosphere may have come about through the depletion of the carbon dioxide by the burial of organic matter. Whenever this happens, as, for example, when the organic matter that formed the coal and oil deposits of the world was buried, the result is a net gain of oxygen to the atmosphere. Whenever organic matter is buried so that it cannot become oxidised and decay, such a gain of oxygen *must* occur.

Planets having a velocity of escape below a certain critical minimum will have lost their atmospheres very rapidly. The Moon, with a velocity of escape of only 1·5 miles per second, and a maximum temperature of 120° C., could retain carbon dioxide and any of the heavier gases. The lighter ones, nitrogen, water-vapour, hydrogen and helium, would escape. In primeval times when the Moon was very much hotter it would have lost even the heavier gases of its atmosphere. For purposes of comparison, I have classed the Moon as a small planet. In the opinion of some authorities, the Earth/Moon system may be classified as a dual planet system rather than that of planet and satellite, the Moon being unduly large in proportion to the Earth.

Mars has a velocity of escape of 3·2 miles per second and a temperature considerably lower than that of the Earth and can, therefore, retain water-vapour and the heavier gases, but not the lighter ones such as hydrogen and helium.

Venus with a velocity of escape of 6·5, not far short of that of the Earth, would be able to retain an extensive atmosphere in spite of her much higher temperature.

We do not know the exact height of the Earth's atmosphere. Meteors or, as they are popularly known, shooting stars, become visible at heights of about 30 to 70 miles. We do not see them until they have become heated to incandescence through friction caused by

THE VELOCITY OF ESCAPE

their rapid passage through the air. The Aurora Borealis is an electrical phenomenon caused by electrified particles entering the atmosphere. The lower portions of the Aurora are usually some 50 to 80 miles up and the highest anything up to 600 miles. At that height the atmosphere must be very rarefied and, although there is no definite point at which it ends abruptly, it gradually becomes more and more rarefied until outer space is reached. We can, therefore, regard its height as approximately 600 miles. For comparison, the Martian atmosphere may have a height of some 200 to 400 miles. If the whole of our atmosphere were at a uniform density equal to that at sea level, it would extend to a height of only five-and-a-half miles. The lowest layer, known as the troposphere, contains nearly ninety per cent. of the substance of the atmosphere and extends to an average height of seven miles although the actual thickness may vary between five and ten miles at different times and places. The total weight of the atmosphere is estimated to be some 6,000 million million tons. The air contained in a room 50 by 30 by 18 feet weighs exactly one ton. It is the weight of air that renders it capable of doing such immense damage when moving at high speed during hurricanes and tornadoes.

The air is constantly being churned up and its ingredients well mixed by winds, storms and the turning of the Earth. If this did not happen the lighter gases would rise and the heavier ones sink so that we should soon be unable to breathe.

CHAPTER VI

The Canals of Mars
Possible Forms of Martian Life—Martian Vegetation —Flying Saucers

THE DIFFICULTIES we have to contend with in examining the surface of Mars are considerable. Owing to the great distance, looking at Mars through a telescope under good conditions is rather like looking at the Moon with the naked eye; for, although we now have telescopes powerful enough to bring the planet to within a tenth of the Moon's distance, such magnifications as they afford do not help us owing to the fact that our atmosphere is not steady enough. Above a certain point, increasing the power of the telescope used, far from showing us the surface of the planet in finer detail, has the opposite effect, the image becoming blurred instead of clearer. Another difficulty is the long interval, fifteen years, between one favourable approach of Mars and the next, due to the eccentric orbit of the planet. The next most favourable approach will occur in 1956.

Even in the very favourable climate of California there may be only two or three days in the year when exceptionally good visibility occurs.

The diameter of Mars is 4,200 miles, rather more than half the size of the Earth. The surface area is accordingly about a quarter that of the Earth's, but since some three-quarters of our planet is covered with

THE CANALS OF MARS

water, and as there are no seas or oceans on Mars, the actual land area is not very much less. The Martian day is approximately half an hour longer than ours, not a very noticeable difference, but the Martian year has 687 days as against our 365, and the seasons there are consequently almost twice as long.

The most brilliant surface markings are the white polar caps, which are seen to grow, and shrink, in the two hemispheres alternately, coming down practically half-way to the equator in winter, and almost disappearing in summer. The fact that the comparatively cool summers of Mars are capable of effecting changes so great that the southern polar cap disappears completely at times, leads us to conclude that the snow can only be a thin covering, certainly nothing like the enormously thick and permanent snow covering at the terrestrial poles.

Next in prominence of the surface features are the red and orange areas which cover the greater part of the planet. They are generally thought to be deserts and, owing to the brilliance of their colouring, it is believed that they may resemble the spectacular and highly coloured deserts of Western America rather than the drab sandy wastes of the Sahara. So far as we can ascertain, they appear to be flat rather than hilly, although it is quite possible that high ground, and even mountains a mile or two in height, may exist. Observation of the south pole shows some evidence of the existence there of mountain ranges.

The most interesting markings, however, are the dark, irregular areas which extend in a belt roughly parallel with the equator, and which were once thought to be seas, indeed, although we now know that they are not, they still retain the original names bestowed on them, such as Mare Serpentis, Mare Sirenium, and so on. The curious feature about these regions is that with the coming of spring and early summer, and the melting of the polar caps, a slowly spreading darkness creeps across them towards the equator, and it is then that the gradually changing colours, from light to dark green and then from yellow and gold to brown, occur. These changes of colour are very much what we should expect to see if we were able to observe our own Earth from space. The cause of these effects here on the Earth would be the seasonal growth and ripening of vegetation. Here these changes would be manifested by the growth of vegetation starting in the temperate latitudes and travelling progressively towards the poles. The tints would deepen gradually. On Earth, the wave of awakening vegetation would travel from equator to pole, while on Mars the movement would be from pole to equator.

The reasons for this difference are that, while the growth and ripening of vegetation are dependent on the rays of the sun and the presence of moisture; on Earth the moisture is always available except in desert regions, while on Mars the moisture must await its liberation by the sun, that is to say, the seasonal melting of the polar caps which provides water without which the vegetation cannot begin to grow. On Mars, where there is at times little or no surface water, growth must commence near the poles where the supply of water first becomes available, then slowly progressing towards the equatorial regions.

Irregular changes in the appearance of the dark areas are noted from year to year and these may well be attributed to the effect of climatic variations on the vegetation. In some years certain areas will receive a

THE CANALS OF MARS

greater share of moisture, while others may suffer from drought. This is exactly what happens here, and if we could view the Earth from a great distance away in space, similar changes would be observed, that is always assuming the terrestrial atmosphere allowed us to get a clear view of the surface, a somewhat doubtful point owing to the thick blanket of cloud and the atmospheric haziness which normally envelops our planet. Climatic conditions on any particular areas of Mars must vary considerably from year to year just as they do on various parts of the Earth.

Now attempts have been made to explain the seasonal changes of colour observed on Mars by attributing them to variations in the degree of hydration of minerals and salts. In other words, it is suggested that the dark areas are covered with soluble salts, similar to the salt pans of aklali plains which exist in some of the desert areas of the Earth, and that these salts absorb moisture from the air when the polar caps begin to melt. The suggestion is that, with the gradual absorption of moisture by the salts, a dark mud is formed, and that this would account for the colour changes observed in the absence of vegetation.

In examining this theory we have to ask ourselves, firstly, why such a progressive dampening of salt or mineral areas should lead to the colour changes observed on Mars. Such colour changes do not occur on the Earth and, secondly, if the changes are not brought about through the agency of vegetation, why they should be seasonal. The presence of moisture might cause a darkening of the surface, but it would certainly not lead to the other colour changes observed.

Even on Earth, where the moisture content of the

atmosphere is incomparably greater than that of Mars, a progressive dampening of mineral areas or salt pans by flooding, light precipitation of atmospheric moisture or heavy showers of rain, could not possibly bring about changes of colouring such as we observe on Mars, and in any case, the dampened areas would quickly dry out again piecemeal and resume their original tint. Another point is that, even if such changes did occur, they would not cover enormous areas with regular seasonal and long enduring effect, were they merely due to the causes suggested.

In view of the arid conditions prevailing on Mars, it is impossible to believe seriously that a mere wetting of surface areas, whatever their composition, would lead to progressive colour changes taking place over enormous portions of the surface. On the other hand, a gradual, progressive spreading of moisture, whether atmospheric or a surface flow over low lying areas, such as the dark portions of Mars (probably ancient sea beds) would bring about just the results we observe if these areas were covered with the type of vegetation that Martian conditions demand.

There is another theory which sets out to explain the colour changes on Mars and that is the volcanic theory of D. B. McLaughlin of the Observatory of the University of Michigan. It is also advanced as an explanation of the Martian canals.

The theory assumes that ash, hurled up by volcanic outbreaks, probably equalling or even exceeding that of Krakatoa, the most violent explosion that has occurred on the Earth during historic times, is carried by the wind and deposited over vast areas, where, either through the action of moisture, or if not this, then chemical reactions between the various constituents of the ash, the colour changes, noted by observers of the planet, are brought about.

It is further suggested that the Martian canals may result from linear chains of small volcanoes situated along major crustal fractures. It is assumed that individual ash from these, each deposit a few miles in extent, might merge in such a way as to produce the illusion of continuity.

The theory first of all rests on the assumption that great volcanic activity is prevalent on Mars. Up to the present we have no evidence of the existence of any volcanic activity there, although it is possible that the yellow mists we sometimes see might result from quantities of very fine ash thrown up into the atmosphere by volcanoes, and then carried by the gentle winds of the planet over great areas.

There is, however, no reason to suppose that such material, deposited over tracts of country, would undergo any such colour changes as are noted. Normally it would impart only a drab or palish hue to the areas covered, or possibly the reddish brown colour of the existing desert areas, or what we assume to be desert areas, since it is not impossible that their colour may be due to vegetation of a reddish hue.

Now since the colour changes progress from poles to equator, in order to establish the first part of the theory, we must assume that most, if not all, of the major volcanoes are sited round the poles, otherwise the progression of the ash deposits would not take place progressively from pole to equator. Such a siting of volcanic chains would appear very unlikely from a geographic point of view.

What is more, violent eruptions, wherever they occur, are intermittent, taking place at rare intervals, years often elapsing between outbursts. This is a characteristic inherent in the very nature of volcanic activity and its root causes. They would certainly never be seasonal or even regular in occurrence.

The theory further requires us to assume that the *Maria* (the dark portions of the planet) are saved from elimination by dust blown over them from the deserts, through their being annually watered with volcanic ash of dark colour. If this were so, it seems curious that the ash should fall only on the dark areas and not on the deserts in such a way as gradually to obliterate the latter.

It seems possible that volcanic ash may be responsible for the yellow fogs, and that it may fall over large areas. It would hardly do so, however, to such an extent as to form surface deposits that would be visible from the Earth. Probably the yellow clouds or veils, as they have sometimes been described, would be no more noticeable to anyone standing on the surface of Mars, than were the immense clouds of very fine dust thrown up from Krakatoa, to the people of, say, India or Australia. The resulting deposit of dust on the ground would probably be even less apparent, being so thinly distributed as to pass, as it did in the case of the Krakatoa explosion, quite unnoticed. It is possible that an observer on Mars, watching the Earth through a powerful telescope for some months after the Krakatoa eruption, might have seen yellow fog-like clouds spreading over large parts of the Earth, even though the people here were quite unaware of their existence, the only apparent effect being exceptionally fine sunsets, and the occasional phenomenon of a coloured sun, due to the abnormal quantity of fine dust present in the atmosphere.

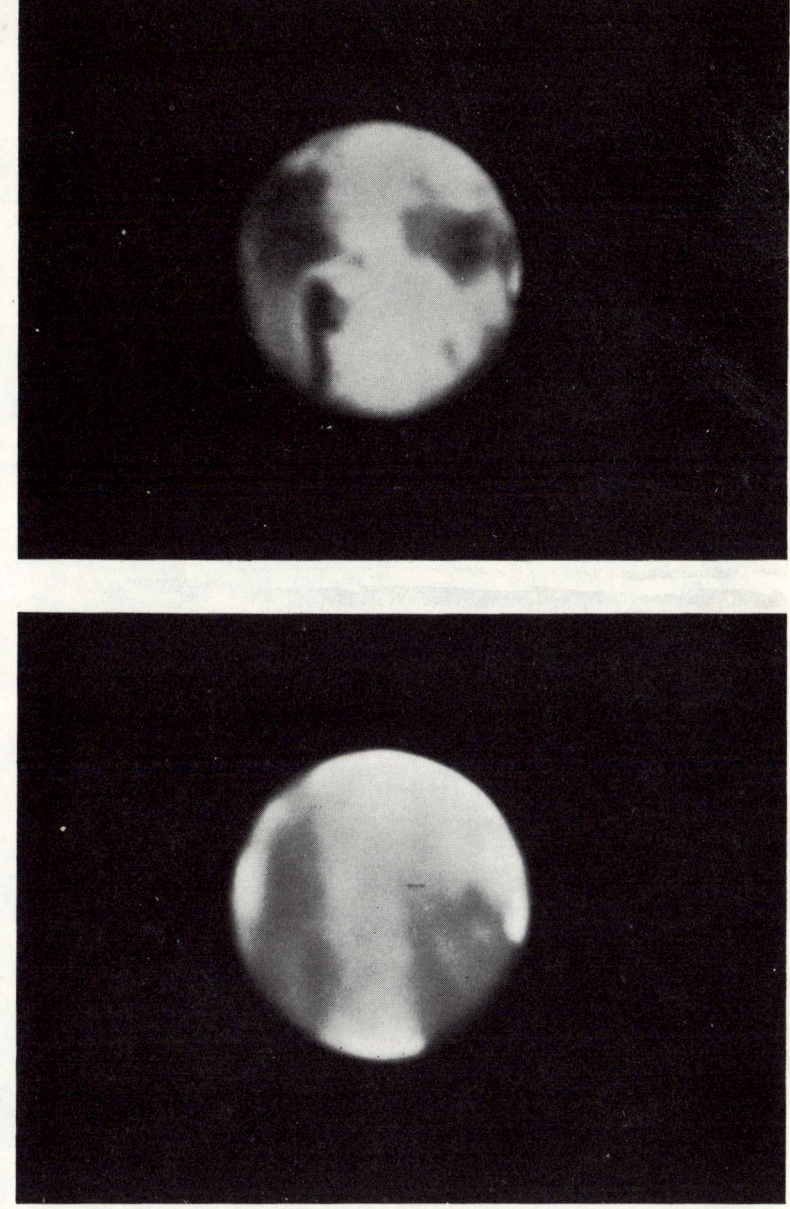

A photograph of Mars taken by G. E. Hale, the blue light (*left*) showing the atmosphere, and the red light (*right*) the surface detail.

Reproduced by courtesy of the Mount Palomar Observatory

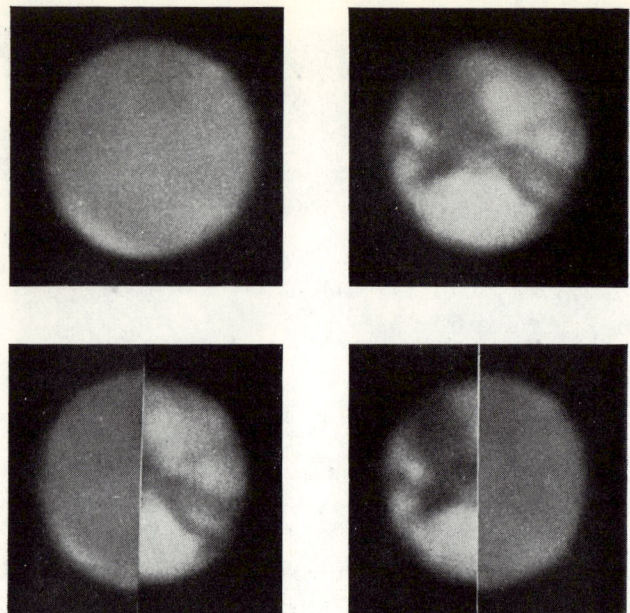

A photograph showing the atmosphere of Mars taken on 2nd November 1926, by Wright. The surface markings are shown in the infra-red photographs but not in the ultra-violet which show the atmosphere only. Hence the difference in size when the photographs are superimposed (*see lower two photographs*).

Reproduced by courtesy of the Mount Palomar Observatory

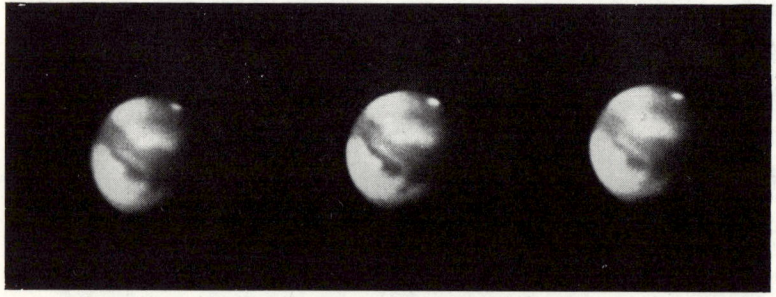

A photograph of Mars taken on 3rd November 1909, by G. E. Hale, showing one of its polar snow caps and some of the surface detail.

Reproduced by courtesy of the Mount Palomar Observatory

THE CANALS OF MARS

The theory sets out to explain the fact that no mountains of any height have been found on Mars, as would be expected if there were any considerable volcanic activity, by assuming that most of the eruptions are of such a violent character as to blast away the cones before they attain any considerable dimensions.

Under this theory the suggestion is made that great volcanic activity might occur if Mars were in the same condition now that the Earth was in the Pre-Cambrian Era, and that in time all this activity will give Mars a structure similar to that of the Earth, and may even lead to the development of oceans, as we believe it did here.

This is to assume that Mars, far from being an old and dying planet, on which mountain building has ceased, and where the surface has been almost levelled by erosion, is actually in the early stages of development and preparation as a home of life. In that case, the vegetation there, far from being among the last remnants of life clinging tenaciously to a dying world, is a vigorous ancestral stock, indicative of the higher forms of life that may be expected to evolve there in time to come.

These then are the theories that set out to explain the seasonal changes of colouring observed on Mars by attributing them to causes other than the perfectly simple and obvious one, the regular seasonal growth, ripening, and decay of vegetation. They would appear to be dependent on so many impossible conditions, incredible occurrences, and unlikely suppositions, as to appear more in the light of attempts to dodge the real issue, the existence of life on other planets, than as serious contributions towards a solution of the Martian problem.

That the green areas cannot be composed merely of

THERE IS LIFE ON MARS

inert matter such as mineral salts or volcanic dust is the opinion of Dr Ernest Opik. His argument is that, if this were so, these green areas would long ago have been buried by the yellow dust storms which occur on Mars, were they not covered by living vegetation having regenerative powers. If it were not for plant life, the whole surface would soon take on the same uniform colouring as the arid desert areas.

One argument formerly advanced against the possibility of the green areas being tracts of vegetation was their appearance in infra-red light. Terrestrial green plants, that is plants containing water and chlorophyll, appear white when photographed through an infra-red filter. It was argued that since the green portions of Mars appeared dark, they could not consist of seed plants or ferns such as we know here. Lately, however, Gerard P. Kuiper, of the Yerkes Observatory in Wisconsin, has proved that *the infra-red spectrum of the green areas compares with that of some of the lower orders of plant life such as lichens and mosses.*

The colour of the surface of Mars provides evidence of the existence of free oxygen, in the past at any rate, and the presence of free oxygen and carbon dioxide makes the presence of vegetation a virtual certainty. This, in association with the colour changes that occur on the surface of the planet, *leave no reasonable doubt of the existence of plant life on Mars.* It may be just lowly forms, mosses and lichens or, possibly, some type of vegetation peculiar to Mars and so totally unknown to us here.

It is impossible at present to say whether animal life or higher forms of life exist there, although we know so little about life or its powers of variation and adaptability that we cannot deny the possibility.

THE CANALS OF MARS

The darker areas may be the beds of ancient seas that once existed and which are now covered with vegetation that springs into seasonal activity under the influence of sunlight and irrigation. The water from the polar caps may spread into them through natural channels. Even a comparatively small area of open water, if such existed, would be visible to us here through the reflection of sunlight from it, but none has ever been observed.

Professor Percival Lowell, who had the advantage of studying Mars under very favourable conditions of both atmosphere and altitude at the Flagstaff Observatory, declared that many of the canals which he claimed to have seen and mapped were double, having the appearance of fine parallel lines some one or two hundred miles apart. He was also of the opinion that the canals, which were faint or even invisible during the spring, became more prominent with the onset of summer and the shrinking of the polar cap and that, while they appeared to grow at the rate of something like fifty miles a day towards the equator, the canals were often observed to extend beyond it into the opposite hemisphere. They would then gradually fade away until, half a year later, taking, of course, the Martian year, they would begin advancing from the opposite polar cap. Lowell also declared that some of the canals would disappear for several years and that new ones would appear where none had previously been seen, and he thought that this might be due to their silting up from time to time and being opened again *or else to the construction of new canals*. At points where two or more canals intersected he noticed darker areas of considerable size covering, perhaps, several thousand square miles and these he called *oases*.

THERE IS LIFE ON MARS

Whatever doubts we may have of the actual existence of the so-called canals, the reality of the darker areas of oases is beyond dispute; *they really are there.*

If the surface of Mars actually were covered with a network of canals as drawn by Lowell and other observers, then his theory of the existence of intelligent beings would be confirmed, but it is only fair to say that recent astronomical opinion is against the reality of the canals as such. The surface of the planet is certainly covered with a mass of fine detail and irregular markings which, it is thought, observers are inclined to connect up as straight lines, owing to their being at the extreme limit of vision.

Atmospheric pressure at the surface of Mars is very low, possibly less than a twelfth its value at sea level here, and we should have to go about eleven miles above the surface of the Earth before finding a pressure equally low; certainly we could not live in such an atmosphere. Despite this low pressure, however, the Martian atmosphere is very deep, a fact accounted for by the weakness of the force of gravity, approximately only one-third that of the Earth's. The fact that clouds have been observed floating more than twenty miles above the surface is proof that there is a far slower falling off of density than we find on Earth. The highest clouds we meet with here are those known as cirrus, usually found at a height of five or six miles. They consist of minute ice crystals, unlike most other clouds which consist of minute drops of water. A somewhat puzzling feature of the Martian atmosphere is the haziness often observed. This might be attributed to fine dust were it not for the difficulty of explaining how so rare an atmosphere could support any quantity of solid particles however fine.

THE CANALS OF MARS

Mars has two very small satellites or moons, Phobos which has a diameter of perhaps fifteen miles and which circles the planet in 7 hours 39 minutes at a mean distance of only 5,800 miles, and Deimos with a diameter of perhaps a little more than seven miles, whose period of revolution is 30 hours 18 minutes. Deimos' mean distance from Mars is 14,600 miles. Viewed from the surface of the planet, Phobos would appear only about a quarter the apparent size of our Moon, while Deimos would probably look merely like a bright star in the sky. As torch-bearers to Mars, they are even less efficient than the Moon is to us; not that this would be any very great handicap as we know by the number of dark nights we experience here without suffering any particular inconvenience.

These miniature moons are of special interest in that one or both of them may become the first bodies, apart from our own satellite, on which space ships of the future may effect a landing. Making contact with them would present little difficulty since their gravitational field is so weak that a man walking on them would bounce up and down at each step he took, and he would only have to leap upwards to jump right away from them into space. Phobos, the nearer of the two, is so close to Mars that it would not be visible from the polar regions, being hidden by the curvature of the Martian globe. It travels round Mars faster than the planet revolves on its axis so that it rises in the west and sets in the east. For a great deal of the time it would be eclipsed by the shadow of the planet. Actually, Phobos revolves in less than one-third of the rotation period of Mars so that it sets only four hours after it has risen. Deimos revolves in little longer than Mars rotates and so remains above the horizon for three days without setting, passing through all its phases twice during that period. A ten stone man would weigh only a few ounces on either Phobos or Deimos, and they can only be seen from the Earth with the help of a very large telescope.

It is not impossible that the canals are, as Lowell held them to be, of artificial construction, and if that were so, then we are forced to conclude that intelligent beings of some sort exist there. If the canals really are the vast engineering works they would have to be in order to carry out the task of conveying water from the polar regions of Mars to the more arid parts of the planet, then the creatures responsible for their design and construction must possess a high degree of intelligence. They might, indeed, be in advance of ourselves in engineering technique. I would make it clear that in all this I am not suggesting that we have any proof of intelligent life on Mars, but merely that it is not an impossibility.

As we have seen, here on Earth, lungs were evolved so as to enable gill-breathing fish and their descendants to become land animals, and it is not beyond the bounds of possibility that on Mars some further evolution of the organs of breathing has come about which would enable creatures living there to breathe in such a rarefied atmosphere.

We, ourselves, are prisoners of our environment, only able to live and breathe at the bottom of a vast sea of air, some six hundred miles deep; just as helpless to escape from our surroundings as fish living in a sea of water are to escape from theirs. As has been recently demonstrated, human beings are able to live and breathe at the top of Mount Everest, a mere 29,000 feet above sea level, only by means of elaborate breathing apparatus.

THE CANALS OF MARS

A lot has been heard of recent years about flying saucers, reports of which have come from responsible sources in many parts of the world. It may be, as I have explained in *Life and the Universe*, that they are natural phenomena of some sort, the exact cause and nature of which has yet to be discovered. *If*, and it is a very big IF, the sensational theory that they are some sort of space ships coming from another planet were true, then there is only one planet they could come from, Mars.

We know that on Mars water is scarce and becoming scarcer as time goes on. Again, if highly intelligent beings do exist there, it is within the bounds of possibility that they have mastered the problems of interplanetary or space flight. Faced with the gradual drying up of their world, the Martians might have considered the possibility of colonising another planet where water is abundant. In this case the only suitable planet, in fact the only one they could reasonably hope to reach, is the Earth. Mars is the only other planet in our solar system on which *intelligent* life might possibly exist. If there is any life at all on Venus it is probably at a very low stage of development.

In two works recently published, it is not only claimed that flying saucers have actually landed, but that, in at least two instances, authors met and conversed by means of sign language, with men from other planets who emerged from the vessels.

One of these beings declared he had come from Venus, the other from Mars. One, the Martian, was actually photographed. Both resembled terrestrial men in all respects: the Venusian is described as being young and handsome, slightly built and with rather long fair hair. The Martian was estimated to have been about six feet tall, with a high forehead and, so far as one can judge, not so good-looking. Apparently, the Venusian was able to breathe quite comfortably in our atmosphere without artificial aid of any sort. The Martian, on the other hand, appears to have had a small tube up his nose. After chatting amicably with the authors, in sign language, both men returned to their saucers which then took off again.

Now, limited as is the knowledge we have regarding conditions on Mars and Venus, we do at least know enough to say with absolute certainty that no creature from either of these planets would be able to live and breathe in our atmosphere for more than a few moments without the aid of elaborate breathing apparatus, nor would they be likely to bear anything but the faintest resemblance to ourselves.

The only conclusion one can come to is that the authors in question were the victims of (*a*) a hoax or practical joke on the part of someone, or (*b*) that they suffered from hallucinations. I should hesitate to think that in a matter like this anyone would be guilty of what Sir Winston Churchill, in one of his inimitable phrases, once described as a "terminological inexactitude."

We are, perhaps, too much inclined to assume that life cannot exist on planets where the atmosphere is very different from ours. The fairly recent discovery that certain forms of life, notably the *anærobic bacteria*, can live without air must to some extent modify our views. Life without air came as a great shock to scientists when it was first discovered.

Some scientists have suggested that anything living on Mars might be quite uneatable for us and, in view of this possibility, astronauts of the future who plan a visit there will have to be sure that their food supplies are ample to last throughout the voyage. Future explorers of Mars will have many risks to face, for one thing the possibility of infection with unknown diseases by bacteria of a kind that have no existence here. Even on Earth new forms of bacteria are evolving and these may in time be responsible for the spreading of new and hitherto unknown epidemics. Some forms of animal and vegetable life, if such exist, could be as dangerous to us as anything known here. There are sound reasons for presuming that, if there is animal life on Mars, it has assumed forms different from anything we know. The probability of this is clear when we consider the great variations in temperature, low atmospheric pressure and dry conditions it would have to stand up to. It seems unlikely that seed plants or ferns of terrestrial type could survive on Mars; for one thing the cold of the Martian nights might freeze the water in such plants. The vegetation that exists may be very different from ours.

When all has been said, it must be admitted that Mars remains a great puzzle to us. Observers of the planet have to contend with great difficulties not the least of which is the unsteadiness of both the terrestrial and Martian atmospheres. Even in such favoured places as Flagstaff and Mount Wilson, periods of perfect seeing may last for only a second or less. Again, when very large telescopes are used the image of the planet is even more blurred than when a lower-powered instrument is employed, for much the same reason that in misty weather low-powered field-glasses give better results than powerful binoculars.

A summary of the various characteristics of Mars is of interest in showing how these resemble or differ from terrestrial conditions.

THERE IS LIFE ON MARS

1. It is now known that the Martian atmosphere contains far more carbon dioxide than that of the Earth. This would favour vegetable life existing there.

2. The polar snow caps shrink in summer and expand in winter. In late spring, when melting becomes rapid, a dark fringe is observed on the outer edge of the polar caps. It could be a polar sea but is more likely to be ground several hundred miles wide which becomes saturated by the melting ice.

3. Dusky areas which have a greenish or brownish hue are clearly seen. These are, generally speaking, permanent in shape and size. They exhibit seasonal changes in colour, visibility and outline, indicative of areas covered with vegetation.

4. The globe of Mars is encircled by a reddish belt in the northern tropics which may be the equivalent of the terrestrial desert belt, Sahara, Gobi, etc.

5. No large open stretches of water, seas or lakes, have ever been observed.

6. So far as we can tell, there are no high mountains, although there may well be some of two or three thousand feet in height. There is some indication of mountain ranges in the polar regions.

7. The atmospheric density has been estimated at about a quarter that found at the summit of Mount Everest, and atmospheric pressure at the surface may be about ten per cent. that of the Earth's. It must be realised that these are merely estimates. We cannot tell definitely as yet what conditions are actually like at the surface of Mars.

8. The planet's atmosphere is considerable with a much slower falling off in density than we find here. This is clear from the fact that clouds have been observed floating twenty miles above the surface. Here

the highest clouds are never more than six miles up. The Martian atmosphere has about 1/40th the water vapour and only a small fraction of the oxygen found here. Moving white patches are observed—probably moist clouds. It is thought that rain, if any, will be limited to light showers. Extensive yellow clouds are seen at times which may be sand or dust storms or, possibly, clouds seen through a yellow atmosphere.

9. Winds are light with speeds, estimated from cloud movements, up to 20 or 25 miles per hour.

10. Observation shows what appear to be mists and hoar frost.

11. The temperature at midday in the tropics is a reasonable one, but the cold at night and during the Martian winter is intense.

12. The rotation period and tilt of axis are similar to those of the Earth.

13. The equatorial diameter of Mars is 4,216 miles, rather more than half the Earth's, and about twice that of the Moon. The volume is 3/20th of that of the Earth with a surface area almost equal to the land area here, since three-quarters of the Earth's surface is covered by water.

14. The mass is only 1/9th and the surface gravity 0.38 of the Earth's. A ten stone man would weigh only about four stone on Mars. The mean density is 7/10ths that of the Earth or nearly four times that of water.

15. The Martian year is considerably longer than ours—687 terrestrial, or 670 Martian days.

16. The lengths of the Martian seasons, taking terrestrial days, are, southern spring or northern autumn 146; southern summer or northern winter 160; southern autumn or northern spring 199; southern

winter or northern summer 182. The southern hemisphere thus has a shorter and warmer summer and a longer, colder winter than the northern. As a consequence of this, the south polar cap melts more rapidly and completely than the northern. It sometimes disappears completely in the late summer but in winter may extend to a latitude circle of 60° diameter, while the northern cap extends to one of 10° less.

CHAPTER VII

The Planet Mars

MARS IS the first planet outside the Earth's orbit and is easily recognisable by its reddish colour. In many ways it is, next to the Earth, the most interesting of all the planets and for a long time now it has been appreciated that conditions there were not unfavourable to the existence of life in some form.

It seems only reasonable to assume that wherever in the Universe conditions are favourable for its existence, life comes into being and begins to evolve; and there is no valid reason why it should not have started on Mars long before it began here, nor why it should not have reached a higher stage of development than anything we know. I am not suggesting for one moment that it has, but merely that there is no reason why it should not have done so. Even if there were no animal life on Mars at the present time, that is not to say that it did not exist at some time in the past. It may have died out through changing conditions, increasing cold, the gradual loss of much of the atmosphere, and the growing scarcity of water. These are conditions that will eventually bring life here to an end for our world is gradually losing its atmosphere, although the rate of loss is a very slow one.

There was at one time some doubt regarding the white polar caps, it being thought that they were not snow at all but some substance other than frozen water vapour, carbon dioxide, for example, or that they might even be an atmospheric phenomenon. Of late years, however, Kuiper of the Yerkes Observatory has been able to prove conclusively that the polar caps of Mars really are snow. The caps, as we have seen, are observed to grow with the oncoming of winter and to shrink with the approach of summer, much as would the polar caps of the Earth, if viewed, say, from Mars.

The atmosphere of the planet is very much rarer than that of the Earth, the Martian atmosphere containing only a small percentage of the water vapour and oxygen found here, although both these gases may have been abundant in the past. It is possible that the ruddy colour of Mars is due to oxidation of the rocks which have extracted most of the oxygen from the atmosphere, and this is one way of explaining its low oxygen content. On the other hand, the loss of atmosphere can be accounted for by the low velocity of escape on Mars, the molecules having gone off into space in the course of ages.

The Martian atmosphere would not be capable of supporting life as we know it here, but life is so adaptable it is within the bounds of possibility that even animal life might have adapted itself to existence in such a rarefied atmosphere. We know, by our experience here, that life can exist in or under the ground, in the air, under water; even in the great depths of the ocean where most forms of life would be crushed out of existence by the immense pressure; or in the oil deposits thousands of feet down. Life can adapt itself to these and other conditions and it is able to assume many and varied forms to suit its environment or to meet changing conditions. It is always possible that a very rarefied atmosphere could support animal life which had developed or evolved to suit existing conditions. *There can now be no doubt that vegetable life of some sort* exists on Mars, and if the presence of vegetable life is admitted it is difficult to deny the existence of animal life, if only because the line of demarcation between animal life in its lower stages and vegetable life is not always easy to define. We can be pretty certain that in the beginning both vegetable and animal life originated from what we call non-living matter; in fact that all life had a common origin.

This is not to say that animal life, if it exists on Mars, resembles anything with which we are familiar here. For reasons connected with temperature and other conditions existing there it has been suggested in some quarters that life on Mars may have developed along, what we may term, insect lines and this is not altogether impossible. Here we are inclined to look on insects as something small and insignificant and we can hardly visualise a dominant form of life formed on such lines. If there were found to be intelligent creatures there which had really constructed vast engineering works (the Martian Canals), always provided these actually exist, then we must presume they are bipeds and possess arms and hands and have large and efficient brains.

It does not follow that life on other planets belonging to our solar system, or to other solar systems throughout the Universe, will necessarily have developed along the lines it happens to have pursued here. It may indeed by very different. Quite unlike anything with which we are familiar or which we can even imagine.

The Martian day has a length almost identical with ours; 24 hours, 37 minutes and 23 seconds to be exact, that being the time taken by the planet to turn on its axis. The axis being tilted at an angle of 20° 10′ as against the Earth's angle of 23° 27′. The seasons are more pronounced than on Earth, that is to say the difference between summer and winter is greater. There is, however, a further cause of the greater variation in the climate of Mars as compared to that of the Earth.

The path pursued by the Earth round the Sun is almost a circle so that the Earth's distance from the Sun does not vary enough to cause any very great difference in the climate. The differences varies by slightly more than three million miles, but in the case of Mars there is a difference amounting to no less than twenty-six million miles. There is consequently a great difference in the climate of Mars when at its nearest and farthest from the Sun and these variations are in addition to the changes in climate that accompany the changing seasons.

The maximum warmth on the planet occurs just before mid-summer, in the southern hemisphere, at the time when Mars is at its nearest to the Sun. The temperature then, in the equatorial zone, ranges from sixty to eighty-seven degrees Fahrenheit or over. That is temperature in the sun and not the shade. Astronauts of the future, landing there from a space ship, might think they had reached a warm and pleasant climate. They would soon be disillusioned. There are neither dense enough clouds nor sufficient atmosphere to trap the Sun's heat, as happens here, with the result that as soon as night sets in the cold becomes intense. It will probably be freezing before sunset and the temperature may well fall to 40 degrees below zero during the night. It is something like the conditions prevailing in the

A photograph of the Moon, Eratosthenes to Plato, taken on
15th September 1919.

Reproduced by courtesy of the Mount Palomar Observatory

A photograph of the Nebula in Andromeda taken on 18th September 1901. It shows what our own nebula would look like if viewed from a great distance away in space.

Reproduced by courtesy of the Mount Palomar Observatory

desert areas of the world only more so. Immediately the sun is down the cold becomes intense.

This is the finest climate we are likely to encounter on Mars. At the poles the temperature may well be more than 100 degrees F. below zero, and when the planet is at its greatest distance from the Sun, it is quite possible that there will be no part of the surface where the temperature is above freezing point. Cold such as this, and the rarity of the atmosphere would make it impossible for most types of life with which we are familiar to exist on Mars.

Photographs taken in different colours provide evidence of an atmosphere at least sixty miles deep. Actually the atmosphere would extend much higher than this in a very rarefied form. Probably, if the Earth's atmosphere were photographed from space under similar conditions, the indications would be that it was perhaps 120 miles deep, although as we know, it extends to at least 600 miles in a very tenuous form. The lowest layer of the Martian atmosphere shows yellow patches of cloud above which is a violet layer which is very effective in absorbing and scattering light of short wave length. Higher still, at estimated heights of five to twenty miles, we find whitish or blue patches which may be fog or cirrus clouds. They are visible in violet light but not in red. Kuiper, using narrow band filters, discovered that the Martian atmosphere is opaque for wave lengths below 4500 A., and almost transparent for waves longer than 5000 A. He found, too, that haze was always least dense on the warmest parts of the surface and that white clouds often formed near the sunset edge, while those near the sunrise edge appeared to break up. This indicates a light scattering atmosphere with a good deal of haze tending to gather in cold conditions and to dissipate in warm.

Photographs of Mars in infra-red light clearly show surface detail while those taken in ultra-violet light show practically none, evidence that the Martian atmosphere is dense enough to scatter the ultra-violet (short wave light) so that it is unable to penetrate to the surface of the planet and out again, as can the longer wave infra-red light. The white clouds show up best in photographs taken in ultra-violet light, while they are hardly seen at all in infra-red photographs, proof that they must be high in the atmosphere. On the other hand, the yellow clouds are seen in infra-red but not in ultra-violet photographs indicating that they are at a low level. Very large areas are at times covered by the yellow clouds which blot out the surface details. They are often very persistent lasting sometimes for weeks on end. In 1909 and 1911 the B.A.A., Mars Section, reported that millions of square miles were covered by these yellow clouds for weeks. Their nature is not definitely established but it is thought they may be dust clouds raised by winds blowing over the desert areas. The dark areas, when near the sunrise limb, are observed to have white edges, believed to be hoar frost.

The amount of solar heat reaching Mars is probably not much more than forty per cent. of that received by the Earth. Tests made by the use of sensitive thermocouples attached to large telescopes have given the following surface temperature readings—dark tropical regions at noon, 68 to 88 F. or higher; bright tropical areas 50 to 68 F.; high latitudes at noon, winter, 32 to 69 F., beneath clear sky and —40 C. under blue cloud. Southern polar temperatures indicate temperatures

THE PLANET MARS

of from 0 to 20 C. in the late summer. It must be appreciated that these are surface temperatures in the sun and not the shade. The winter temperature of the polar caps has been calculated as low as —70 C. or approximately 125 degrees of frost F. It is believed, however, that this may be temperature above the clouds covering the poles, and should this be the case, the surface temperature would be considerably higher. At midsummer the temperature at the poles is appreciably above freezing point.

One rather curious feature about the polar caps is that they show up most clearly on photographs taken in light of short rather than long wave length. If the caps were merely a surface detail, we should expect the opposite to be the case. It would seem, therefore that they consist of masses of snow and ice beneath a permanent bank of cloud, probably high cirrus of no very great thickness so that light of long wave is able to penetrate it.

It is possible that water is carried from the melting poles towards the equator in the form of vapour and not in the liquid form. The great seasonal variations in temperature might be mitigated for animal life, if such exists, by migration from one hemisphere to the other. This would be relatively easy owing to the small size of the planet and the fact that there are apparently no geographical barriers to movement.

CHAPTER VIII

The Great Mystery of the Martian Canals

THE SO-CALLED canals of Mars were first discovered by the famous Italian astronomer Schiaparelli in 1877. Using a low-powered telescope he found a number of narrow streaks or fine lines that crossed the surface of the planet connecting up larger dark areas. He described these lines as " canali," an Italian word meaning a canal or channel of water such, for example, as the canals of Venice and not necessarily signifying as does the word, *canal* in English, an artificial channel made by human labour.

The word canal having been used, people immediately jumped to the conclusion that they were actual channels or canals of artificial construction, and that there must be intelligent beings on Mars who were responsible for their construction, and considerable interest and excitement was aroused as a result of the discovery.

In 1894, Percival Lowell, the distinguished American astronomer built Flagstaff Observatory in Arizona for the special study of Mars and other planets, and after long observation came to the conclusion that the lines were too straight, too regular and too geometrical in pattern to be of anything but artificial origin. He believed them to be an immense and wonderful irrigation system built by the Martians to pump and convey water from the poles to the waterless interior in order to grow crops and stave off death by thirst and starvation. There are no open seas or oceans on Mars and water there must certainly be scarce. A growing shortage of water and its eventual disappearance from the surface is a natural and inevitable process on any planet. It will happen here eventually and, provided the human race is still in existence when the time comes, those then living will certainly use all the means at their disposal to utilise every available source of supply.

Lowell thought that on Mars highly civilised creatures were waging a desperate struggle for existence in the face of the gradual drying up of their world and their own eventual extinction. To-day, although few, if any, astronomers would agree with Lowell's opinion, no actual disproof of his theories has ever been obtained. The fine dark lines on the planet's surface may be seen; at times covering much of the area with a network of lines, dots and streaks. Some of these are as much as 3,000 miles in length and from fifteen to twenty-five miles wide. Recent opinion is that the markings may not be actually as straight and regular as they appear to be and that they are possibly of natural and not artificial origin. We should like to know more about them for neither large modern telescopes nor photography have so far been able to solve the mystery. If it were ever proved that the canals were of artificial construction, we should have irrefutable proof that Mars was inhabited by a form of life as high or higher than ourselves. On the other hand, if these are natural features of the planet's surface, it would be interesting to know just what they can be for, so far as we know, there is nothing like them anywhere else.

Lowell believed that what we see are not the actual waterways, which at so great a distance would be quite invisible, but broad belts of vegetation spreading out on either side of the channels, the irrigated areas. So, too, he thought the dots represented large irrigation centres at certain points or junctions, perhaps near to or adjoining the sites of Martian cities.

Now, quite apart from anything else, the fact that areas of what appear to be vegetation can be seen from so great a distance shows that, (*a*) the areas must be vast, and (*b*) the vegetation, for the colour changes to be observed at all, must be well marked. The vegetation of Mars may be something in the nature of lichens, if only for the reason that these can survive in conditions of low temperature and shortage of water such as would be fatal to the continued existence of most terrestrial plants.

It is, however, not easy to believe that the growth of lichens would be sufficiently pronounced in either volume or colouring to permit of such changes as are seen to occur being observable from the Earth. It seems more likely that the changes of colouring result from vast areas of irrigated land on which some type of crops are growing. They may well be species of plants peculiar to Mars and totally unknown to us here and for which, consequently, we have no names. It is also difficult to give a reason why vegetation of any sort should grow and ripen over vast areas, and they must indeed be vast for us to see them at all, unless there is some form of irrigation, either *natural* or *artificial*. Conditions on Mars must be arid and it is not easy to say why large areas of the planet should benefit from natural irrigation.

Why, for example, should water flow to certain areas,

unless they happened to be pure jungle or tropical swamp forests, neither of which are likely to exist on Mars. Even if such types of vegetation did exist, their prevailing colour would be an unchanging green, just as they are here. Conditions on Mars are almost certainly rigorous and unsuited to the growth of any kind of tropical or sub-tropical plants but that is not to say that crops of a nature suited to their environment and which have adapted themselves to the conditions prevailing there, may not thrive. After all, even here, a visitor from the tropics or sub-tropics might wonder how life survives at all under the winter conditions prevailing over large parts of Europe, Asia and North America. Low temperatures and hard living conditions, however, produce vigorous stocks, whether animal or vegetable and the human animal is no exception.

We do not know to what extent life is capable of adapting itself to existing or changing conditions, although here on Earth we have pretty good evidence of its adaptability, for it has taken on many forms in the course of its evolution. What could be more different than a man, an octopus, a butterfly, a dogfish or a cabbage, yet they are all merely different forms of life, all built up from exactly the same materials and actually consisting mainly of water. In the light of modern discovery we can only assume that wherever in the Universe prevailing conditions do not render it impossible, *life will come into existence*, develop and adapt itself to suit its environment.

With our present knowledge, and the instruments so far available, it is impossible to say whether any life at all exists on Venus, the planet nearest to us and, apart from the Earth and Mars, the only other planet in our particular solar system on which it seems life would not be absolutely impossible. When we examine the atmosphere of Venus, it is only the *upper part* that is available to us. The lower portion is out of the reach of our instruments and there is, therefore, an inclination to say that the atmosphere of Venus is such that no life can exist since there is no trace of water vapour or oxygen in the upper atmosphere. There may be plenty of both lower down, next to the solid surface, and it is the condition of the lowest strata of an atmosphere that decides whether life can or cannot exist in it. If we take the Earth's atmosphere as an example, we find that the upper regions are quite different from the lower. What is known as the Troposphere, the lowest layer of all, only reaches to about seven miles above the surface, yet it contains practically the whole of the water vapour in the atmosphere, and this in spite of the fact that the latter extends to a total height of not less than 500 and quite possibly as much as 700 miles. Eighty miles up the Earth's atmosphere may be almost pure hydrogen with neither oxygen nor nitrogen present.

If we can imagine a party of astronomers on Mars, equipped with all the latest types of instruments and telescopes that we have, examining the Earth, it is safe to say that they would be quite unable to ascertain whether animal life exists here. The suggestion has been made that they might infer the existence of intelligent life from the glow of large cities on the dark side of our planet. This seems very doubtful. Even if the light were visible at such a distance and through our hazy atmosphere, they could hardly be expected to infer the cause. The only thing they could be certain about is the existence here of vegetation.

MYSTERY OF THE MARTIAN CANALS

That Mars still has a very considerable atmosphere is shown by photographs taken of the planet in ultra-violet and infra-red light; the former showing a larger image due to the fact that it includes the atmosphere, while in the infra-red photograph, only the solid body of the planet appears. Photographs of Mars, taken recently with the world's largest telescope at Mount Palomar, show white areas at the poles which it is thought may be banks of fog lying above masses of snow. Other photographs show the large dark areas previously referred to.

When in 1877 Schiaparelli turned his new eight-and-three-quarter inch aperture refractor on Mars, which was then at favourable opposition, with the object of determining accurately the Martian latitudes and longitudes of a number of points on the planet's surface, he experienced exceptional seeing conditions and discovered an extraordinary amount of surface detail that had not previously been suspected. He also found that the existing maps of Mars were inaccurate and decided to make an entirely new one with a nomenclature based partly on the ancient names of Mediterranean and Middle East lands and seas, and partly on names in Biblical and classical mythology. To-day, his Martian maps and names, with additions made by Lowell and Antoniadi, are standard.

Schiaparelli, following established custom of previous map makers, called the dark areas *mare* (sea), *sinus* (gulf or bay), *fretum* (strait), *lacus* (lake), according to size and shape. He discovered a number of narrow dusky streaks which crossed the light areas in various directions and these he called *canali*, meaning channels. The word was translated *canals*, implying that they were of artificial origin, a meaning that Schiaparelli never intended to convey. Although a small number of the larger and darker canals had been mapped by earlier observers, the discovery of a number of new and fainter ones caused considerable excitement and astonishment as well as a good deal of scepticism. This was increased when, in 1879 and 1881, he continued his observations and drew up new maps of the canals showing them straighter, narrower and more geometrical. Some he even drew as double, like two railway lines running parallel across the surface of the planet.

The canal theory found little support among astronomers in Europe, but in America, in 1894, Lowell founded an observatory at Flagstaff, Arizona, for the observation of planets, particularly Mars. He used apertures of eighteen inches and twenty-four inches and he and his assistants discovered a great number of canals, so that presently their maps were covered with a network of geometrical lines that appeared to radiate from the polar caps, intersecting one another at darkish spots which Lowell named *oases*. Their straightness, narrowness and frequent intersections, seemed to discount the possibility that they might be ravines or river beds and so Lowell declared them to be canals or watercourses, constructed by the Martians for the purpose of irrigating the vegetation growing on either side of them with water from the melting snow at the poles. His theory met with considerable opposition and to-day finds little support.

Antoniadi and other observers, using large telescopes, declared many of the canals to be merely unconnected lines of dusky spots, edges to shaded areas, or irregular diffuse shadings. On the other hand, it has been pointed out by G. de Vaucouleurs that the canals mapped by Schiaparelli, whether or not they are broken or continuous formations, are still seen in the same places, enjoy a seasonal cycle and develop into prominent dark bands lasting for years and then becoming faint again. Many of them, too, obey the laws of perspective as actual markings would. Dr. Barnard, a very experienced observer, having the advantage of the sixty inch reflecting telescope at Mount Wilson, a far more powerful instrument than anything Lowell had at his disposal, was of the opinion that the canals are broad, diffuse and irregular in outline and not the sharp, clearly defined, geometrical network which Schiaparelli and Lowell held them to be.

How, then, can we reconcile these very divergent views?

For one thing, acuity of vision is important and this may vary with different observers. There is also a tendency for the eye to connect up irregular markings which are at or near the extreme limit of vision and glimpsed only momentarily. Periods of perfect seeing, even in the clear atmosphere of Arizona or California, may last only for a second or so and that, incidentally, is why it is so difficult to obtain satisfactory photographs.

When Mars is most favourably placed for observation, a telescope with a power of seventy-five brings it to the same apparent size as the Moon seen with the naked eye. The smallness of the image, even in a large telescope, makes it impossible to obtain photographs showing the minute detail of the surface markings. The planet is not bright enough to permit of its being photographed by instantaneous exposure, while, with time exposures, the unsteadiness of the atmosphere tends to blur the image so that fine detail is lost. The limitations on what a photograph can reveal are further

THERE IS LIFE ON MARS

increased by the fact that the surface detail of Mars is so intricate that much of it is finer than the grain of a photographic plate. In spite of these difficulties, some excellent photographs of the planet have been taken and some of them do actually show canals. It is possible that, with the close approach of Mars in 1956, using cine-photography in conjunction with the 200 inch Mount Palomar telescope, photographs might be taken with exposures of one-tenth of a second or even less. If so, the structure of some of the canals may be clearly shown.

CHAPTER IX

The Planet Venus

Possibility of Life There—Possible Life on the Moon

VENUS REVOLVES around the sun in an orbit between that of Mercury and the Earth. The planet's mean distance from the Sun is just over 67 million miles and the orbit is so nearly circular that the distance varies by less than a million miles. As a result, the orbital speed remains fairly constant at around twenty-two miles per second with little variation, while the light and heat received from the sun are about twice that reaching the Earth. At superior conjunction the planet's distance from the Earth is about 160 million miles, and at inferior conjunction, 26 million miles. With the exception of the Moon, a few of the smaller asteroids and an occasional comet, no other bodies in space come so close to us. The diameter is approximately 7,700 miles, only 220 miles less than that of the Earth. The mass is about 0·82, density about 0·89 and surface gravity about 0·86 those of the Earth. The velocity of escape is very slightly less than that of the Earth.

What we call twilight here on the Earth, a gradual merging of daylight to darkness, is due to the atmosphere which scatters the sunlight. If the Earth had no atmosphere, darkness would come with the suddenness of switching off an electric light immediately the sun had set. Twilight occurs on Venus, as we can observe, when the planet is between us and the Sun, appearing as a crescent when the tips of the horns extend far round the circumference of the dark limb, unlike those of the crescent Moon where they appear as the two ends of a diameter. The effect seen in the Venusian crescent is proof that there is a region of twilight for, although the Sun is not shining directly on it, the dark portion of the planet is still visible through the sunlight scattered by the Venusian atmosphere.

When a transit of Venus occurs and the planet is just impinging upon, or just leaving the disc of the Sun, the edge of the planet which is outside the Sun is surrounded with a bright line of light. This phenomenon is due to the scattering of sunlight by the Venusian atmosphere. Transits of Venus are of somewhat rare occurrence. This is because the orbit of Venus is inclined to that of the Earth at an angle of $3\frac{1}{2}$. If both the orbits were in the same plane, Venus would be seen to pass in front of the Sun each time it changed from an evening to a morning star. As things are, we only see it projected on the disc of the Sun when it is near one of the two points where its orbit crosses the plane of the orbit of the Earth. The last transits were seen on 9th December 1874 and 6th December 1882, and the next two will take place on 8th June 2004, and 6th June 2012.

Venus is the only planet, apart from Mercury and perhaps Pluto, that has no satellite and since the location of the polar axis is not known, it is impossible to say whether seasons occur. If they do, they would be very much shorter than ours, the Venusian year having only 225 of our days. If there are seas or oceans they would have very considerable tides, for the Sun would have a much greater effect on them than has the Moon on terrestrial seas.

Venus is sometimes spoken of as the Earth's twin sister if only because her size, density and mass are very close to those of the Earth. We know, however, that conditions on Venus are different from those that prevail here. Except for the Sun and the Moon, she is the brightest object in the sky and, as we have seen, at her nearest approach comes within twenty-six million miles of us. That being so, it might be thought that we could find out quite a lot about her. Actually, we know very little.

The reason for this lies in the extremely dense and permanent atmosphere of the planet and the thick mantle of cloud that always envelops her. So far, the atmosphere has prevented all attempts by astronomers to penetrate it and find out what the solid surface of the planet is like. Being nearer the Sun and, therefore much hotter, we should expect her to have, if anything, rather less atmosphere than the Earth, instead of a much denser one with far more cloud. No surface markings have been detected and without these it is impossible to be certain of the rotation period. Some authorities consider this may be as much as thirty days or thereabouts and, if this is so, since the length of the Venusian year is only 225 of our days, there will only be seven or eight days in the year. It is a curious state of affairs and, although the latest opinion is that the day there may not be less than three or four of our weeks, the problem remains unsolved. The four weeks' rotation period, however, is merely an estimate and by no means certain. In 1912, Sir Norman Lockyer estimated the period as being $23\frac{1}{4}$ hours, and Belopolsky's work with the spectroscope showed evidence of a rapid rotation. We shall not definitely know the truth until we are able to get direct evidence, and that will not be until some means of penetrating the Venusian atmosphere is invented as, no doubt, it will be in time. It is fairly safe to say that the temperature evidence is definitely against an enormously long rotation period. If Venus does not rotate, or rotates very slowly, then we should expect to find the side turned to the Sun very much hotter than it is and, conversely, the night side very much colder. The clouds that prevent us from seeing the solid surface of the planet would make the stars invisible to any creatures living there, if such exist, and it is unlikely that direct sunlight ever reaches the surface, so that even at noon the only light is perhaps a faint kind of twilight.

A curious thing about the Venusian clouds is that they do not appear to be composed of water vapour, no trace of water or oxygen having been detected in the atmosphere. The only gas discovered is carbon dioxide. It has been estimated that the atmosphere of Venus above the clouds contains a quantity of carbon dioxide equal to a layer two miles thick on the Earth's surface. We can only speculate what the atmosphere is like at or near the surface of the planet. If carbon dioxide is plentiful below the clouds and vegetable life exists, it would liberate free oxygen from the carbon dioxide which it absorbs, *and if there is free oxygen there may also be animal life.* One theory in regard to the clouds of Venus is that they may be clouds of dust swept up by tremendous wind storms raging continuously, immense tornadoes caused by the great variation of temperatures between the night and day sides of the planet due to a slow rate of rotation. The objection to this theory is, that if it were so, we should

THE PLANET VENUS

expect the clouds to be brown or, at least, a dirty grey instead of being, as they are, a dazzling white.

The fact that there is no trace of oxygen or water vapour in the upper atmosphere is sometimes advanced as an argument in favour of there being no life on Venus. There may be, however, plenty of both these gases in the lower part of the atmosphere and it is the condition of the lowest layer that decides whether or not the atmosphere of a planet is capable of supporting life. The air we breathe consists principally of four parts of nitrogen gas to one part of oxygen, with less than one per cent. of argon and minute quantities of several other gases.

It is sometimes said that Venus may be a world of oceans and vast swamps, very much like our world was hundreds of millions of years ago, possibly inhabited by primitive forms of life; or again, that it may be a planet in such an early stage of development that, as yet, life has had no chance to come into existence. Normally one would expect a planet of the size and position of Venus in the solar system to be at least contemporary with, if not ahead of, the Earth in evolutionary development. Such development, supposing it had occurred, would not necessarily have followed anything like the lines it has done here. As in the case of Mars, living creatures, if any do exist, may be quite different from anything with which we are familiar. Very different indeed to anything which we can even imagine, for, given different conditions and environment, life could develop along many different lines and adopt an infinite variety of forms. If we want proof of this we have only to look around us and see what evolution has accomplished here on one small planet out of the limitless numbers that probably exist throughout the Universe.

Venus, being one third nearer the Sun than we are, should be about twice as warm, since light and heat increase inversely as the square of the distance. Twice as near would mean four times as warm. As it happens, these estimates agree with actual measured temperatures. If we take the average temperature of the Earth as being sixty degrees F., that of Venus will probably be around one hundred and twenty. Even though we might find the tropical parts of the planet rather too hot for us, we should probably be quite comfortable in the polar or temperate regions and, in any case, if there is life there, it will be adapted to living in the higher temperature.

In the opinion of some astronomers, Venus is a lifeless world, her whole surface a desert where cyclones and tornadoes, immeasurably more violent than anything we experience here, rage continuously, whirling great masses of dust into the atmosphere, so that the planet is enveloped in perpetual obscurity. They base this view on the fact that water vapour cannot be detected in the atmosphere and that the great difference of temperature between the night and day sides of the planet would give rise to violent winds, while the greenhouse effect of the exceptional amount of carbon dioxide present in the atmosphere might mean that the surface temperature during the day is above the boiling point of water.

Failure to detect water vapour would be understandable if the Venusian clouds or haze extended to a height of five miles or so above the surface. On Earth there is very little water vapour five miles up. It is somewhat surprising that conditions on a planet so nearly comparable to the Earth in size should differ so greatly from those existing here. Venus is not so

THE PLANET VENUS

much nearer to the Sun as to make life impossible if other conditions besides temperature were favourable. Some authorities have ruled out all likelihood of life in any form existing there. Others go so far as to admit the possibility that certain forms of bacteria may live in the Venusian atmosphere. There is little doubt that, with the continued advance of science, we shall presently learn much more about conditions on Venus. Meantime, we can only wonder and surmise.

We must not lose sight of the possibility that the atmosphere of Venus, like that of many other planets, may be quite unbreathable. It is also likely that many planets throughout the Universe have atmospheres in which the creatures existing on them are able to breathe but in which we should quickly perish. There is also the possibility, although it does not seem very likely, that the surface of Venus is entirely covered by water, just as the Earth probably was at one time. Even now, if the whole surface of the Earth were levelled out, it would be covered with water to a depth of one and a half miles and there would be no dry land at all. It is only the immensely large and deep basin of the Pacific Ocean that prevents the world being all sea. In that case, life here could never have advanced beyond the fish stage.

One theory to explain the existence of the Pacific Basin, and there are plenty of sound arguments to back it, is that at a period when the Earth was still in an intensely heated and plastic condition, the material that filled what is now the bed of the Pacific Ocean, was torn away to become the Moon. The reason such a mass could have been torn out of the Earth is understandable when we realise that the gravitational pull of the Sun must have raised enormous tides in the plastic material of the Earth long before the oceans existed, and as the world was then spinning with much greater velocity, this, combined with what is known to physicists as the *force of resonance*, caused the solar tides to rise to enormous heights until at last they became too great for stability and a large mass of the intensely hot and semi-liquid surface was hurled into space. Upholders of this theory argue that the volume of a body the size of the Moon would almost exactly fill the bed of the Pacific Ocean. Again, the bed of the Pacific, according to some geo-physicists, consists of the same substance as the Earth's middle layer, basalt, while the thin layer of granite which forms the bed of all the other oceans, is absent. The argument is that it was torn away when the Moon was formed. Then again, the average density of the Moon is only 3·3 as compared to 5·5 for the Earth, which leads us to infer that the Moon is composed of basalt and granite, the materials of the Earth's outer layers, and that she contains none of the heavy iron core of the Earth's centre. Whether the theory is right or wrong we cannot say definitely but the fact remains that the existing relationship of land and water here is entirely fortuitous. The world might just as easily have been *all* instead of, as it is, *only three-quarters* ocean.

Even the Moon may not be an entirely lifeless world where nothing ever happens. This is suggested by certain changes that appear to have taken place in the shape of some of the craters of recent years. It is thought that these may be due to large meteorites striking the walls of the craters or the mountain tops and, also, possibly to moon-quakes or movement of the surface through volcanic action.

Some astronomers are now watching for evidence of

meteorites striking the lunar disc, since the impact of a large one might show up as a tiny flash on the dark portion of the Moon.

It was recently reported from Russia that scientists there had discovered evidence of a very tenuous atmosphere surrounding the Moon, said to be approximately 1/20,000th that of the Earth's. If this were so, we might expect to see meteorite trails above the lunar surface. Some astronomers have reported seeing such trails. More recent observations, however, have failed to confirm evidence of any atmosphere.

Of all the craters of the Moon, perhaps the most interesting is Eratosthenes; if only because of the curious dusky markings that are seen spreading up the slopes of the mountain and over the walls of the crater. They change their shape from night to night and they *may be some low type of vegetation peculiar to the Moon.*

If this is so and they are vegetation of some sort, a kind of fungus perhaps, it would, as in the case of Mars, be proof that life evolves on any of the bodies in space where conditions make its existence possible. It seems a much more likely thesis than to assume that it occurs only on one planet, the Earth, and nowhere else in the Universe.

The great plain known as the Mare Crisium, is subject to heavy mists or opaque clouds. These cannot be mist or fog as we understand them since there is neither air nor water on the Moon. Their nature is not yet known but they may consist of carbon dioxide, one of the gases given off in volcanic eruptions, an indication that active volcanoes still exist on the Moon. The Mare Crisium or Sea of Crises, is not, of course, a sea in the terrestrial sense. Like the other so-called seas of the Moon it is merely a vast plain, with here and

THERE IS LIFE ON MARS

there ridges and craters spread over its surface. The Moon's gravity is one-sixth that of the Earth's, so that a man could jump six times as high or throw a stone six times as far as he can here. For the same reason volcanic activity there would be far more violent than that which occurs on the Earth.

Another crater on the Moon which shows what may be signs of life is that of Plato. A darkening of the crater is noticed from night to night which may be vegetation of a sort. There are probably other portions of the lunar surface on which similar phenomena occur but this has not yet been confirmed.

It is difficult, however, to credit that any sort of life exists on the Moon. The surface is alternately roasted during the long day, and frozen during the equally long night by a cold far surpassing anything known on the Earth. The possible signs of life that have been mentioned, are, according to some astronomers, due to variation in the appearance of the surface detail caused by changing altitude of the sun. Others maintain that they must be due to some strange form of vegetation.

The Moon displays other mysterious features. There are the curious bright streaks surrounding Tycho and Copernicus. Many theories have been advanced to explain them. They may be due to salts oozing from the ground, or perhaps a large meteor that once struck the Moon and splashed molten material around, or they may be great cracks filled with white lava. There is, too, a mountain called Piton which rises to 8,000 feet, the peak of which glitters and sends out rays like those of a searchlight. The reason for this is not known.

CHAPTER X

Life on Other Worlds

On Mars—On Venus

So FAR we have been considering the possibilities of life occurring on Mars and other planets throughout the Universe, under conditions with which we are familiar, and following the assumption that similar conditions are essential wherever life is to gain a hold. It is clear that life, as we know it, could only exist on bodies in space where conditions are more or less comparable with those existing here.

If, however, we are to assume that life is capable of adapting itself to widely differing conditions, to an extent beyond anything within the bounds of our experience, then the area of its possible occurrence is enormously extended, and the scope of its potential development on planets, so apparently inhospitable as Mars, correspondingly increased.

If, for example, forms of life exist which have as their basic atom, that of silicon, instead of carbon, as is the case here, then creatures so constituted would be able to live and thrive in temperatures and under conditions, that would be fatal to us. Again, if there is anywhere in the Universe an active system of converting the energy of a sun or star into the energy of living creatures, which is more direct and efficient than the somewhat involved and indirect way in which this end is accomplished here, then forms of life may occur of which we can form no idea. Such life would follow laws and principles entirely beyond our present comprehension.

Throughout the Universe there is a fundamental uniformity in the construction of matter. The same elements exist everywhere and all matter is built up from them. Everywhere matter obeys the same laws. It seems likely, therefore, that wherever in the Universe life exists, the carbon atom forms its basis. From this we must assume that living matter can only occur under somewhat limited and specialised conditions. Life would certainly not be possible in any form on a star, for the temperature of the stars is such that only the simplest of chemical compounds can exist in them. Nor can we reasonably expect it to occur on planets which are either too hot or too cold, or on planets attached to stars which vary much in their output of radiation, or on planets belonging to multiple star systems, since the temperature variation would be too great. For one star with a planet or planets on which life exists, there may be many others with planets that never had, and never will have, life of any sort on them.

Intelligent life is likely to be a rarity in the Universe at any particular period in time, since it occurs, as on the Earth, for only a fractional part of the life span of a planet.

It is only on planetary bodies, which are really nothing more than small, dead and decaying suns, that any life is possible. Not until surface decay is well advanced, can life, at least the higher forms of life, come into existence or make much progress.

Life is so adaptable and its determination to survive so strong that we are not justified in assuming that it could not succeed in gaining a hold on many of the planetary bodies in space. Certainly, the exceptional combination of circumstances favourable to life existing on the Earth are found on no other body in *our solar system*. At the same time, it is going against all reason to imagine that a chemical reaction, at once so all pervading and so adaptable, should be confined to only one of the planets in our particular part of space.

Even with the best of modern instruments it is impossible for us to be certain what conditions are like at the surface of a planet such as Mars. Definite information will probably have to await the landing there of space explorers of the future. Only they will be able to tell us, for example, if the canals actually exist and, if so, what they really are.

If there were intelligent beings on Mars it is an interesting flight of fancy to speculate on their probable appearance. Invariably the fiction writers picture them as small replicas of ourselves (the little Martian men) or as great, clumsy creatures of the most bizarre appearance, something like a cross between an octopus, an ant and an elephant. Of one thing we can be fairly certain; they would be small and active, with large and efficient brains, walking upright and possessing arms and hands. It is most unlikely that they would resemble us in anything but the most remote degree. Certainly they would be nothing like men in miniature.

Widely differing conditions and environment would have a marked effect on evolutionary trends. Had terrestrial conditions changed materially during the evolution of our remote ancestors, we should certainly not be what we are to-day. We might have been more, or less intelligent than we are, and our physical appearance would probably have been quite different.

Doubtless, among the countless billions of stars in

the Universe, there are many having a planet or planets revolving round them on which conditions approximate fairly closely to those existing here and on which life has come into being and evolved along very much the same lines it has followed on the Earth. The same process has probably occurred in the past on countless planets which have long ceased to exist, and will, perhaps, occur over and over again on planets, the material of which now exists only as interstellar gas, or even cosmic energy, yet to be converted into matter.

When the first space travellers from the Earth set foot on Mars it will be interesting to know what they discover. Even though they, themselves, may never return, they may be able to relay the information back to Earth.* The probability is that they will find themselves on a strange and barren world, where the only visible life is some form of lichens covering areas that were once the beds of ancient seas, surrounded by tracts of red and yellow desert; or they may discover that the same areas bear curious forms of vegetation, unknown to us. In that case, even the deserts may be strangely exotic and highly coloured, with cactus-like plants adapted to their environment, plants which serve to bind and hold the dust of the deserts or, even as Dr Opik has suggested, feed on it.

If, and I would again emphasise that, in the light of our present knowledge, this is a pure flight of fancy, our space travellers found that the Martian canals really do exist as the artificial construction that Professor Lowell and others held them to be, then their adventures on Mars will be exciting to say the least, for they will be in contact with living creatures as intelligent or perhaps even more intelligent than themselves. What

* By short wave radio.

such beings would look like is anybody's guess, and how they would behave towards their terrestrial visitors would remain to be seen.

Once it got there, landing a space ship on Mars would present no particular difficulties. It would be easier than landing on the Moon if only because the Martian surface is far less rugged and broken than that of our satellite. A landing on Venus would be a very different proposition. We have no idea what the atmosphere of Venus is like or what the clouds that envelop her really consist of. In the not very distant future, means may be discovered of penetrating the Venusian atmosphere and finding out what the surface is like. As we have seen, some authorities believe that it is a vast desert from which the dust is continually hurled up by tremendous hurricanes and tornadoes that rage unceasingly. They base this assumption on the great difference in temperature between the night and day sides of the planet, which would give rise to windstorms, far more violent than the worst we experience here. According to this theory, the Venusian clouds consist of dust which is carried up into the atmosphere and given no chance to settle.

If this is so, it would be extremely hazardous to attempt a landing on Venus. The space ship would, in all probability, be wrecked before it succeeded in reaching the surface, and even if it did get safely through the turbulent atmosphere, it would be in grave danger of being smashed up in landing, through the inability of the crew to see where they were going. True they might be helped to some extent by radar, but this would not enable them to tell whether or not they were landing in dangerously broken country. If the clouds of Venus are just clouds of dust, the atmosphere would, in all probability, be worse than a pea-soup fog, with visibility practically nil. Until such time as we can discover much more about the actual conditions prevailing there, any attempt to land would appear to be outside the bounds of reasonable possibility, even when we have space ships that can travel to the planets.

Possibly, in the not so distant future, scientists will be able to analyse the Venusian atmosphere in its lower, as well as its upper regions, and to devise some means of penetrating the cloud layers and finding out what the surface of the planet is really like, its actual speed of rotation and so on. It may be that we shall then find conditions quite different from those described. Of one thing, however, we can be certain; that with so dense a cloud cover, the surface must be wrapped in almost impenetrable gloom. If living creatures exist, they have probably never seen the sun or the stars.

CHAPTER XI

Inter-Planetary Travel

and Space Ships of the Future

MANY PEOPLE are inclined to regard the prospect of inter-planetary travel and space exploration of the future as akin to science-fiction, and dismiss the whole subject as something belonging to the realms of phantasy. Those who doubt the practicability of space flight, however, might well consider what would have been said in 1900, or even in 1903, when the Wright Brothers made their first brief flights in a heavier-than-air machine, had anyone suggested that, long before the half century was out, aeroplanes would be flying the Atlantic in a few hours.

There are many problems to be solved before space travel can become a reality, but I should hesitate to say that any of them will prove insuperable. Firstly, there is the question of the means of propulsion. Aeroplanes, whether of the conventional airscrew type or jet-propelled, cannot operate at a height of more than about fifteen miles owing to the rarity of the atmosphere at higher altitudes, while balloons carrying light instruments are unable to rise much above twenty-five miles, for at this height the density of the surrounding air is little greater than that of the hydrogen in the envelope. At the present time, the only known means of propulsion that will operate in a vacuum, or near vacuum, is that of the rocket. With the future development of atomic power a more effective and economical means of propelling space ships may be evolved, but for the present the rocket is enough.

It is believed* that the governments of more than one country have now authorised investigation into, if they have not already set on foot, active preparation for the building of a type of small multiple-rocket propelled missile, which would be shot outwards into space, and then into an orbit around the Earth, with a velocity sufficient to keep it there for ever after as a miniature satellite or space station.

The object would be quite a small affair, packed with scientific instruments which would relay information back to Earth by short wave radio transmission. The messages would continue to come in until such time as the batteries were exhausted, after which they would, of course, cease, although the object would continue to orbit the Earth just as does the Moon, and for exactly the same reason. This would be a first step in establishing space stations. Valuable data regarding conditions in outer space, temperature, atmospheric density, the effects of infra-red and cosmic radiation, etc., would no doubt be obtained. Up to the present, the fastest man-made rockets travel at a speed of some 5,000 miles per hour, and the greatest distance so far attained by such a missile outwards from the Earth is about 250 miles, at which distance gravitational pull is still about ninety per cent. of that operating at sea level.

The gravitational pull of the Earth extends outwards in space to infinity, slowly diminishing with increasing distance, so that at a distance of something like a million miles it becomes so slight as to be almost negligible so far as its effect on a space ship would be con-

* Now confirmed.

cerned. At a distance of 12,000 miles a one pound weight would weigh merely an ounce. In space there is no such thing as up or down but simply movement outwards, away from a body such as the Earth, or inwards, towards it.

Once outer space has been reached, there is practically no limit to the speed which could be safely attained by a space ship, but during its progress through the atmosphere this would have to be limited, otherwise friction with the air would quickly heat it to incandescence. Even in outer space acceleration would have to be gradual. Were it too sudden, the effects might be fatal to the crew.

The first step in inter-planetary flight will probably be the sending of unmanned rockets to the Moon. They would contain automatic cameras and other instruments, and would be remotely controlled from the Earth, being made to circle the Moon at fairly close range and then brought back. Before this can be done, however, it may be necessary to establish what are known as space stations at varying distances outside the atmosphere. These would be virtually small artificial satellites circling the Earth at speeds varying with their distance. Once set going in their orbits by rocket propulsion, at the calculated speed required, they would continue to orbit the Earth without the need for any further power. The pull of gravity would hold them for ever after, just as it does the Moon or any other of the bodies in space. They would be built up gradually from materials carried out to them by rocket ships.

In order to overcome the force of gravity and escape from the Earth, a rocket or space ship would have to attain a speed of slightly over 25,000 miles an hour.

THERE IS LIFE ON MARS

For such a vessel to circle the Earth in a close orbit, without using further power, once the initial speed had been attained, it would have to attain a velocity of approximately 18,000 miles an hour. Obviously such speeds would have to be reached by stages.

In travelling to the Moon or Mars, more than one type of space ship would probably be employed. There would be (1) the space stations circling in permanent orbits round the Earth, the Moon or Mars. There would also be (2) the rocket ships which would carry materials and supplies from the Earth to the space stations. Then (3) a type of ship which might never land anywhere, but would travel from an orbit round the Earth, to one round the Moon or Mars, as the case might be. The rocket or ferry ships would carry supplies, materials and personnel to them from the Earth, or the space stations, and would land these from them, on the Earth, the Moon or Mars as required.

Rocket power would be used not only for propulsion but for braking, when approaching a planetary atmosphere, or landing on an airless world such as the Moon, or one with a very thin atmosphere like that of Mars, and for steering or changing course, which would be carried out by discharging rockets at an angle to the line of flight.

The No. (3) ships would be shifted from an orbit round the Earth, to one round the Moon or Mars, or vice versa, by increasing speed to overcome gravity and so moving outwards in a gigantic spiral. No very great power would be required for this, and the rocket motors could be comparatively small and light as compared to those of the No. (2) ships. Yet another type of ship, which we will call No. (4), would carry out swift and direct journeys from the space stations to the Moon or Mars. They would be equipped with very powerful rocket motors and would attain enormous speeds.

Members of the crews of the various types of ships would wear pressurised space suits when leaving them for re-fuelling, building up space stations and so on. They would float in free space, attached to the parent ship by lines, to prevent any risk of their floating right away in space, and they would propel themselves by impulse from small portable rocket-firing devices. It will be realised that in space everything is weightless, and this fact would make constructional work on space stations comparatively easy, since all the materials would remain in position without visible means of support. Space station, rocket ship, men and materials would all be travelling in free orbit at a uniform speed.

Should it become possible to employ atomic energy for propulsion, the whole problem of space travel may be greatly simplified and accelerated. It might then be possible to eliminate any intermediate stages and to travel direct from the Earth to the Moon, or to the planets.

A later development would be the establishment of bases on the Moon or Mars, in the form of pressurised buildings, possibly underground, in which the astronauts could live in some sort of comfort.

Another problem arising is that of pressure inside the cabin of a space ship. Normally the atmospheric pressure we live in is one of fifteen pounds to the square inch, at sea level, or about a ton on every square foot of our bodies. The reason we can withstand such a pressure is, of course, that the air inside us presses outwards with equal force. In space, the ship would be in a vacuum, and the cabin would have to be strong enough to withstand the enormous pressure outwards of the air contained in it. The human body is capable of adapting itself, giving sufficient time, to existence in an atmosphere about one-third that of the normal value.

Normally air consists of four-fifths nitrogen and one-fifth oxygen, the oxygen therefore, contributing only three pounds pressure towards the total atmospheric pressure of fifteen pounds to the square inch. It is possible to live for a considerable period in an atmosphere of pure oxygen, and this being so, the problem of maintaining an atmosphere of reduced pressure, not only in the cabins of space ships, but in the space suits worn by the astronauts, is greatly simplified. The difficulty of removing carbon dioxide produced by respiration could be overcome by using sodium peroxide, which not only removes carbon dioxide but replaces it with fresh oxygen. Supplies of pure oxygen can be stored in the liquid state in suitable containers. A man performing moderate physical exertion requires a little over three pounds of oxygen per day. When resting or sleeping, however, the amount needed is reduced to about a pound. Excess water in the atmosphere could be removed by chemical means or by condensation.

Some kind of forced ventilation would also be necessary to remove waste gases as soon as they were formed. Normally the gases we exhale being warmer and, therefore, lighter than the surrounding air, tend to rise, so that the fresh air we breathe in is continually taking their place. Were it not for this we should quickly suffocate. This form of air circulation is due to differences of weight caused by gravity; but in free space, where there is no gravity, it would not operate.

For the same reason there would be no up or down in the cabin of a space ship, and the crew could sit, or

INTER-PLANETARY TRAVEL

stand, just as easily on the floor, the walls or the ceiling, or even float about in mid air.

There are two ways in which this lack of gravity could be overcome. The first is by maintaining a gradual acceleration during the first part of the voyage, or until past the point at which the gravitational pull of the Moon, for example, exceeds that of the Earth; and then a gradual deceleration. This method will remain an impossibility until such time as we can devise a new and more efficient means of propulsion than any yet available, nuclear energy perhaps.

The second involves the use of centrifugal force by causing the airship to spin. It has been calculated that a rate of spin of one revolution every three seconds for a globe twenty feet in diameter would be sufficient. The crew would not be aware of this spinning and would have the sensation of normal weight, being able to walk all round the globe without suffering any inconvenience. Incidentally, a space ship need not be of the conventional cigar shape usually pictured. In outer space its shape would be immaterial. It could be globular, or square, for example. A globe would probably be the most practical form, and the necessary rate of spin could be imparted by using the steering rocket apparatus in a short burst. Once imparted, the spin would continue until the end of the voyage, when it could be stopped by a second burst of power just before landing.

A rotating globe would have another advantage, the maintenance of a constant temperature in the ship during flight. If it were not rotating, one side, that exposed to the sun, would become excessively hot, and that on the opposite, or night side, extremely cold. Some means of heating would have to be carried for use when flying on the night side of a planet where the cold would be intense.

Another problem is the possible effect of radiation on the ship and its crew when outside the protective covering of the atmosphere. Solar radiation might present no great obstacle. The dangerous ultra-violet rays would not penetrate the hull of the ship, and certain types of glass are known which, while opaque to ultra-violet, will allow normal light to pass through.

Cosmic rays, on the other hand, can penetrate considerable thicknesses of lead. Their origin and composition are not yet understood. They appear to be charged particles, possibly protons, travelling at speeds approaching that of light. Although the atmosphere acts as a shield to some extent, we are constantly bombarded by these rays. The intensity of the radiation increases with altitude, until, about twelve miles up, it approaches fifty times the intensity at sea level. This is the maximum. At still higher altitudes the radiation decreases until it is from fifteen to twenty times that operative at the Earth's surface.

What the effects of this radiation would be on crews of space ships over a prolonged period is not known, but it is on record that in 1935 Stevens and Anderson, in the balloon Explorer II, spent several hours in or above the region of maximum intensity without suffering any ill effects, and several Russian balloonists appear to have done the same in recent years. It would seem from this that the danger of cosmic rays may have been exaggerated.

Another danger that would threaten the safety of a space ship is collision with a meteor or a meteorite. It has been calculated that more than 700,000,000,000 of these missiles enter the Earth's atmosphere every

INTER-PLANETARY TRAVEL

twenty-four hours. Most of them are minute particles, far smaller than a grain of sand. Four or five million are probably less than an eighth of an inch in diameter (meteors), while only five to ten are large enough to reach the surface (meteorites). Very occasionally a really large one, weighing from several hundredweights to many tons, lands on the Earth or falls into the sea. The great majority of all sizes fall in the sea simply because most of the Earth's surface is water.

What renders even a small meteor dangerous is the velocity, which may exceed 150,000 miles an hour. Even so, the chances of a space ship being struck by one is very slight, probably in the region of 10,000 to one against during a journey to the Moon. The answer would probably be self-sealing walls on the same principle as the bullet-proof petrol tanks used in aircraft. Collision with a large meteorite, an inch or more in diameter, would spell disaster, but the chances against this are probably hundreds or even thousands of millions to one against.

The question of supplies of food, water and oxygen raises no great difficulty. On polar expeditions the food allowance per man is a daily ration of two pounds. For astronauts, who would be making little or no physical exertion, one pound per day might suffice, with an equal allowance of water, and two pounds of oxygen. A great deal of the water required could probably be reclaimed from the atmosphere in the cabin by distillation and purification.

The problem of the effects of acceleration on the body in taking off from the Earth and reaching escape velocity are not insuperable. Escape velocity could be reached in nine minutes, at an acceleration equal to two gravities, or three and a half at five. Given adequate protection, men can withstand greater accelerations than these for longer periods without injury. Such accelerations would not, however, be necessary in actual practice. A space ship could gain velocity much more gradually by going into an orbit at a height of fifty miles or so, and then using its motors to build up speed in horizontal flight. Once orbital velocity had been attained, it could either shut off the motors and wait to be refuelled, or continue acceleration until escape velocity was reached.

The question of space navigation, although a complex one, should not present any great difficulty to the astronauts of the future. One advantage they would have over terrestrial navigators is that the stars would be visible to them at all times. The navigator of a space ship would be able to fix his position by observations of the Sun and planets, with the help of an almanac containing tables giving the position of the planets at any time. For this he would employ a sextant or whatever variation of this instrument might be developed for astronautical observations. He would also be able to calculate the distance of the ship from any planet it might be approaching, by the apparent size of the body, until close enough for a small radar set to take over this task.

Velocity would be difficult to calculate accurately, but it has been suggested that this might be accomplished by sending out a given number of pips per second from a radio station on the Earth. If the ship were moving towards the Earth it would receive more than the given number of pips per second, if away from it, less.

Communication between the Earth and the ship might also be effected by radio. Normally the heavi-

INTER-PLANETARY TRAVEL

side and other ionised layers above the Earth stop the penetration of radio waves having a length of about ten metres or over. Were it not for this we should be unable to receive radio programmes. As it is the radio waves are bounced back to Earth when they meet these ionised layers. Radio waves, shorter than ten metres, are able to penetrate the layers and so could be used for inter-planetary communication.

In the case of the Moon, quite small short wave radio transmitters would suffice to keep up communication with the Earth stations. Communication between the Earth and Mars would be more difficult owing to the enormously increased distance, but even this could be done with the assistance of the wire mesh structures used in some types of radar. These are merely radio mirrors which collect incoming messages, or focus outgoing ones.

There would be an inevitable time lag in receiving messages. It takes $2\frac{1}{2}$ seconds for a signal to travel to the Moon and back at the speed of light, 186,283 miles per second. Even when at its closest to the Earth, it would be nine minutes before a reply could be received from Mars. The words of a message would be received on Earth more than four minutes after they had been spoken on Mars. Another means of communication that might be employed is that of light waves for sending messages in morse, using powerful searchlights as transmitters, and mirrors, and sensitive photo-electric cells as receivers. It should be within the bounds of possibility to send such messages over millions of miles.

In this brief review we have examined some of the main problems which will have to be overcome before inter-planetary travel becomes a reality, and the lines along which a solution may be found have been indicated. There are many other important questions which will have to be resolved before voyages outside the atmosphere can be undertaken. That they will be resolved in time there is little doubt. For those who are interested in the subject a number of very interesting works have recently been published which deal fully with the many problems of space travel and planetary exploration.

There would be no particular object in getting to the Moon or the planets unless the explorers did something when they got there and the next chapter is concerned with some of the problems to be faced once landing has been effected.

CHAPTER XII

Landing on the Moon

On Mars—The Lunar Bases—The Martian Base

THE MOON is generally considered to be a completely airless world and, if that is so, the first precaution astronauts landing there will have to take is that of protecting themselves against the risk of being struck by meteorites. The airship itself would have to stand in the open wherever it landed and take its chance just as it would have done during the journey through space. The crew, when out of the ship, would wear space suits having oxygen containers, some means of heating and ventilation and also protection against rapid alternations of extreme heat and intense cold. A suit that was merely a sort of glorified diving dress would not suffice. It would have to be built to withstand the enormous outward pressure of the atmosphere it contained. Probably it would be constructed of metal with jointed limbs, electrically operated, and controlled by the wearer. The only times at which space men would be safe from meteorites would be while in a cave or under some overhanging rock which would act as a shield, so perhaps they would just have to accept the risk as an occupational hazard.

It has sometimes been claimed that the Moon possesses a very tenuous atmosphere, one ten-thousandth or less at the surface than that of the Earth's. If this were so, owing to the lunar gravity, the density of the atmosphere would fall off with altitude far more slowly than does that of the Earth and, consequently, even such a rarefied atmosphere might afford considerable protection from meteorites. It may be remembered that these missiles become incandescent in their passage through the upper reaches of the terrestrial atmosphere where the air is probably even more rarefied than the most perfect vacuum that can be produced in the laboratory. We cannot count on this, however, for the chances are that the Moon has no atmosphere at all.

One argument in favour of the existence of an atmosphere of some sort is the appearance of lunar mists which have been observed from time to time. They may, however, be accounted for by volcanic activity still taking place. We cannot definitely say that there is no atmosphere, nor can we be sure that water does not exist in the form of hoar frost, or ice, in caves, where the temperature would be constant, and far below freezing point. It has been suggested that the brilliance of some of the mountain peaks when the Sun strikes them may be due to their being capped with ice. Water could not exist in the liquid state owing to the lack of atmospheric pressure. If it does exist in the form of ice, it would be available to explorers equipped with electrical heating apparatus, the only form of heating they could employ on the Moon. In the future, no doubt, nuclear energy will supply all the heating and power required.

The temperature of the lunar surface has been accurately measured by means of the themocouple. At midday the temperature of the surface rocks is a little above that of boiling water (212° F.), falling at night to 250° below zero F. Owing to the absence of air, there is an enormous variation in temperature

LANDING ON THE MOON

between sun and shade. An explorer standing in the sun at 212° F., would, by merely moving into the shade, find himself in a temperature many degrees below zero F.

It will be realised that the great variations in temperature mentioned apply only to the actual surface. Below ground, only a few inches down, they would be greatly reduced owing to the insulating quality of the lunar rocks and the volcanic ash, that we have reason to believe covers much of the surface. Some authorities are of the opinion, based on radio measurement, that this ash or dust averages about a millimetre in thickness. There may be, however, places where it has accumulated to a considerable depth and this would form one of the hazards to be taken into account when attempting to land a space ship.

Oxygen is not likely to be found on the Moon in the free state, but since the Earth's crust consists of more than fifty per cent. oxygen by weight, we may assume that the same holds good of the lunar crust. If this is so, it might be possible to extract it in quantity.

No doubt at a fairly early stage in lunar exploration some attempt will be made to construct a pressurised building of some sort, possibly below the ground, or inside a cave. Such a building would give the explorers much more roomy accommodation than the space ship could afford. If a telescope of even moderate power could be erected on the Moon it would prove of inestimable benefit to astronomy. Seeing conditions would be perfect and the riddle of the Martian canals would probably be solved at once. The establishment of a lunar base would also provide us with a natural space station on the road to the planets.

If water exists in the frozen state, in caves, as it well may, two of the greatest problems are solved at once for the water could be electrolysed to produce oxygen. Even the rocks might be made to yield water, for water is a constituent of many minerals and can be extracted by the application of heat. During the day heat could be obtained by focusing the rays of the sun, or by the use of electrical energy, at any time. The water might even be used to provide the fuel for chemical rockets (hydrogen and oxygen) for propelling space ships, but probably, by the time of which we are speaking, an effective form of atomic propellant will be available, a few pounds of which would carry the largest vessel to Mars and back.

It might even be possible eventually to produce food on the Moon by means of hydroponic, namely soil-less cultivation, carried on inside pressurised buildings, probably some type of glass houses of special construction. All pressurised buildings would have to be fitted with air locks at their entrances.

Transport on the lunar surface might be effected by means of track vehicles equipped with pressurised cabins and driven by rocket motors, electric power, or, what is more likely by the time men succeed in reaching the Moon, atomic power. The surface area of the Moon is twelve million square miles, about equal to that of the African continent, so that it would be a long time before explorers from the Earth would be able to cover more than a small portion of it.

Astronauts landing on Mars would find their task much simplified by comparison with the difficulties that they would face in establishing a lunar base. For one thing, the Martian atmosphere, although a rarefied one, would no doubt be deep enough to prevent meteorites, apart from those of considerable size, from reaching the surface, just as is the case here.

LANDING ON THE MOON

The existence of an atmosphere would greatly simplify the construction of buildings. They could, for example, be much lighter and less strongly built than those on the Moon since there would be some compensating atmospheric pressure on the outside. Even simple domes of some plastic material supported by the air pressure inside them might suffice. Whether it would be possible to move about on the surface of Mars wearing only breathing equipment and a comparatively light type of clothing is not yet certain.

Track vehicles with pressurised cabins, or even wheeled ones driven by atomic powered motors, could be used for exploration over distances. It might even be possible to fly aeroplanes of the conventional type in the Martian atmosphere, although they would be dependent for propulsion on rocket or atomic-powered motors. Owing to the comparative flatness of the Martian surface they would be able to keep near the ground and so take advantage of greatest air density. The force of gravity having only a third of its terrestrial value would, to a great extent, compensate for the rarity of the atmosphere.

Water, we know, exists on Mars, and oxygen is most likely present in quantity although combined with other elements. To obtain free oxygen and so provide a breathable atmosphere would seem to be a matter of simple chemical engineering. In the opinion of some nuclear scientists, it may indeed become possible, with the mastery of nuclear energy, *to change the entire atmosphere of a planet and to raise or lower its temperature by many degrees; or even to alter its orbit, bringing it nearer to, or moving it farther from the Sun.*

Fantastic as all this may sound, I should hesitate to dismiss it as a mere pipe dream. The science of nuclear physics is, as yet, only in its infancy and no one can say what almost unlimited power to create, or to destroy, may come with its future development.

All this, however, lies in the very distant future, but the establishment of a Martian base may come about within a measurable distance of time, perhaps not more than a century or two. Once the plastic domes had been constructed, pressurised buildings would be erected inside them, so that, if any accident happened to the domes themselves, the inhabitants of the buildings would be safe; and since it is a scientific fact that a man who has been breathing an atmosphere of pure oxygen can live for at least ten minutes without air, anyone caught outside the buildings could be brought into one of them in time to save his life. The domes would be inter-communicating by means of air locks, so that an accident to one would not affect the others. Each of the domes would also have air locks leading to the outside. The temperature inside the buildings could be maintained at any required level both by day and night.

Tools and machines would be brought by space ship ready for assembling, but local materials would be used for the buildings. There would be a chemical plant where ores and minerals would be brought in for treatment for the extraction of oxygen and other materials. Robot machines, operated by remote control, would probably do most of the outside work, road making, excavating, haulage and so on.

There would be a plant for processing food from vegetation grown in special pressurised buildings, and possibly even from the vegetation native to Mars.

The first base, and any others constructed later, would be established in the tropical areas so as to take the full advantage of the best climatic conditions available.

LANDING ON THE MOON

As we have seen, the establishment of a Martian base may come about within the space of a century or so, but nowadays scientific progress is so rapid that the time factor may be materially shortened. New discoveries now follow one another with startling rapidity. The establishment of a space station or artificial satellite may come about within the next ten years or so, or it might occur even sooner. It may not be long before we are able to send multiple rockets carrying automatic cameras and scientific instruments on a controlled voyage to the Moon and back. It will be a step on the way to sending manned rockets or, perhaps, atomic-powered space ships on the same journey. Men may succeed in reaching the Moon within the next half century and it will then only be a matter of time before they are able to reach Mars.

It has been said that space navigation would become very difficult once we had passed beyond the orbit of the Moon, but I think the difficulties have been exaggerated. Very accurate calculations can be made of the exact position of the planets at any time in the near or distant future. Even for a voyage to the Moon accurate calculations would have to be made, otherwise a space ship or rocket might cross the orbit of the Moon when that body was still hundreds of thousands of miles distant. Even so, the crew of a space ship would still be able to rectify errors by a change of direction, steering their ship towards the Moon, or, in the case of Mars, towards the planet, so as to meet it coming round in its orbit. Certainly it would be difficult to calculate speed, but even this problem may be overcome by the instruments of the future. By calculations, for example, of the increasing size of whichever body in space the ship was approaching, or the decreasing size of the body it was leaving, over a given period. There might also be the system of sending out a series of radio pips from the Earth at a given rate per second. To say that the navigation problems of space travel cannot be overcome is surely to underrate the inventive genius of our scientists. No doubt the same thing was said in times past of those who suggested that a means might be found of navigating the oceans of the world. The means *were* found, and the instruments to make it possible *were* invented. It would be no great surprise to hear at any time during the next few years that scientists had succeeded in establishing a space station outside the Earth's atmosphere, or that they had despatched an unmanned rocket to the Moon, so rapid has the progress of science now become.

And supposing, just for argument's sake, the first men to land on Mars should find that intelligent life exists there, and not merely low forms of vegetation. Admittedly it is most unlikely, although not impossible. After all, the Martians, if they exist at all, and have no telescopes more powerful than those we possess, would be unable to tell whether intelligent life exists on the Earth.

If, however, we take a flight of fancy and assume that intelligent creatures exist, what would be their reactions to the landing of human beings in a space ship from Earth?

Presumably much the same as our reactions would be to the landing of a space ship from Mars. If the Martians were fortunate enough to land in an area where people of reasonable intelligence could arrive in time to prevent their destruction by the stupid or ignorant, they would be safeguarded as objects of the highest scientific interest.

LANDING ON THE MOON

No doubt the reactions of intelligent Martians, if such exist, would be roughly similar, but how they would react to further developments is another matter. Any changing of the Martian atmosphere by people who had mastered control of nuclear energy, on the scale that has been suggested, would be fatal to life on Mars, at any rate, to all higher forms. Some low forms of life, bacterial or vegetable, might survive the change. Some indeed might even thrive under the altered conditions, so much so as to become a plague, much as did rabbits and prickly pear when introduced to the climatic and other conditions of the Australian continent.

All this is, of course, supposition, and we must assume that, when they succeed in getting there, men will find that Mars is a world inhabited only by low forms of vegetation and probably bacterial life of a sort. A world where they will have free scope, within the limitations imposed by existing conditions, to explore and to carry out scientific experiments. A world which will be theirs to develop and, perhaps in time, who knows, to colonise.

Note.—It has been said that causing a space ship to rotate in order to give the crew a sensation of gravity, would render it impossible for them to navigate in space. Admittedly this would be a problem, but I should hesitate to say it will be beyond the capacity of scientists of the future to solve it.

Incidentally, we do not yet know what effect a lack of gravity would have on human beings, nor do we know what the effect of prolonged exposure to cosmic rays outside the atmosphere would be.

Protection against ultra-violet rays could no doubt be devised, but cosmic rays penetrate everything except a thickness of lead impossible to provide.

The atmosphere of Mars might give some protection against an excess of ultra-violet rays, as does that of the Earth, but this would not hold good for the airless Moon. Even the Martian atmosphere might not provide a very effective shield.

Notes

(1) Quite recently, some scientists and astronomers have come to the conclusion that the temperature conditions on Mars may not be nearly so rigorous as we have hitherto believed. It seems probable that readings of the surface temperature may have been incorrect, and the climate may therefore be appreciably warmer than these would indicate. The large amount of carbon dioxide present in the atmosphere should have, as it were, a greenhouse effect, just as it has here and on Venus, by trapping the warmth of the solar rays, and so raising the temperature of the planet above the level at which it might normally be expected to stand. If this opinion is correct, conditions favourable to the vegetable life of Mars, and higher forms of life, if such exist, would be materially increased.

(2) Most scientists are now definitely of the opinion that life arises spontaneously from inorganic matter when conditions favourable to its generation are present.
A theory put forward by I. A. Oparin, the Russian born scientist, throws some light on the way in which hydrocarbons may have been produced from carbides, which resulted from chemical reactions in the hot damp atmosphere of the primal Earth as it cooled down. He suggests that other reactions occurring between hydrocarbons and ammonia produced certain nitrogen derivatives. These in turn reacted with each other, and with water, to produce organic chemical compounds, among which were the proteins. Over a very long period, probably numbered in hundreds of millions of years, some of these developed the ability to absorb chemicals which enabled them to maintain and reproduce themselves.

The possibility of this is borne out by experiments conducted by S. Miller at the University of Chicago, who, as a result of exposing methane, ammonia, hydrogen and water, which are thought to have been the principal components of the Earth's early atmosphere, to an electric spark continuously for a period of eight days managed to produce three of the amino acids which form the constituents of proteins. The proteins in turn compose a large part of all living matter.

(3) Dr Wendell Stanley of the University of California, considers that it is difficult, if not impossible, to draw a dividing line that separates sharply the organic from the inorganic, that is to say, living from non-living substances. It is a question of gradual increase in the structural complexity of matter.

When we come to consider the question of LIFE we have to realise that there is no hard and fast line dividing living from non-living substances. The fundamental difference between a lump of stone, say, and an ant is mainly one of greater complexity of matter. The fact that the ant is what we call *alive*, and the stone is not, is merely one example of the different behaviour of matter under widely differing conditions and circumstances.

Hitherto it has always been believed that life originated on the Earth. Now it is thought possible that life had its origins in the material from which the Earth and other planets were built up.

As Hoyle has pointed out in his monumental work, " Frontiers of Astronomy ", in the days when it was believed the Earth and other planets were formed in a completely molten condition, a pre-planetary origin of life would have been an impossibility, since the very high temperatures engendered would have destroyed instantly any complex organic molecules.

Now, however, that it is considered the planets were probably aggregated from an accumulation of cold bodies, our views must be modified accordingly.

NOTES

It is probable that, under the influence of ultra-violet light from the Sun, a combination of such relatively simple substances as water, methane, ammonia and hydrogen, may be built up into molecules of considerable complexity, containing perhaps 20 to 30 separate atoms, such as the amino acids. Such molecules would contain large stores of internal energy supplied by the ultra-violet light of the Sun.

Now the normal rule is that molecules charged with internal stores of energy would have a tendency towards certain chemical changes that would tend to rid them of their accumulated energy. There would be, as Hoyle puts it, a break back into the original materials. If, however, the molecules remain at a reasonably low temperature, they are forced to dispose of their energy by aggregating themselves together into molecules of increasingly greater complexity, relatively small quantities of energy being released at each stage of their metamorphosis.

So far as the origin of life is concerned the question is, how did smaller molecules of the right type manage to come in contact with one another. According to Bernal, favourable conditions for this consummation would probably occur if the molecules were coated as a film on the surface of a solid particle. Such a condition would be more likely to occur before the planets became aggregated as a whole, and while the materials from which they were ultimately built up were still distributed as a mass of small independent bodies.

This is probably a normal process in the evolution of matter that occurs at every formation of a planetary system; the very earliest stages in the production of living from non-living matter.

It is unlikely, as Hoyle says, that animals and plants, as we know them, originated in inter-planetary space. The vital preliminary steps, however, that resulted in the production of life may have occurred there.

Very recent theories and discoveries in what may be termed the field of Planetary Astronomy go to show that solar systems may be far more numerous than has hitherto been supposed.

THERE IS LIFE ON MARS

It is now thought that the majority of stars may be accompanied by a retinue of planets just as is our particular star, the Sun. The natural corollary is that, far from being a rare phenomenon as has hitherto been presumed, life is probably abundant in the Universe. The chances are that almost any planet on which conditions are in any way suitable carries life of some sort. There are no doubt innumerable planets throughout the Universe on which intelligent life exists, on which it existed in the past, and on which it will exist in the future.

This doubtless holds good for innumerable billions of planets that have long ceased to exist, and for countless others that have not yet even come into existence, planets that will not be formed for billions of years to come. I am here taking the English measure of a billion as a million millions. According to Hoyle there may be at least a hundred thousand million solar systems similar to ours in the Galaxy, and the Galaxy, it should be remembered is only one of countless other similar aggregations of stars in limitless space. Within range of the great 200 inch telescope at Mount Palomar there are possibly upwards of a thousand million galaxies similar to ours, and the range of this telescope only touches as it were an ultra microscopic fringe of infinity.

(4) ÅNGSTRÖM UNIT. Å.Ū. Tenth-metre, 10^{-10} metre, $\frac{1}{10,000}$ micron. The Å.Ū. is a unit of length, used particularly for measuring wave-lengths of light.

MICRON, μ. One millionth of a metre, 10,000 Ångström units.

Additional Notes
Radio Astronomy

Ever since, with the invention of the telescope, astronomy became a serious science, the frontiers of space have been receding with every increase in the power of the instruments employed. With the 100-inch telescope, objects five hundred million light years distant became visible, and now, as mentioned earlier, the 200-inch Palomar telescope reveals distant nebulae or galaxies, the light from which has been travelling for a thousand million years, at a speed of six million million miles a year, before reaching us.

In the next year or two, when Radio Astronomy gets into its stride, it will be able to cover an area of space at least five, and possibly ten times greater than that now within range of the Mount Palomar telescope.

Up till about two years after the termination of the 1914-1918 war, it was thought that our galaxy represented the whole extent of the universe. Dr. Edwin Hubble, to whose memory this book is dedicated, was the first to discover the existence of what are known as the extra-galactic nebulae. It was he too who later found that the distance of the great nebula in Andromeda, and of some other galaxies, had been greatly underestimated. The Andromeda nebula, for example, was found to be not about 700,000 light years distant, but twice as far away; and consequently, instead of being far smaller than our own galaxy, it now proves to be somewhat larger.

Martian Canals

Last year (1954), when the National Geographic-Lowell Observatory Expedition visited South Africa under the leadership of Dr. E. C. Slipher, one of America's leading

THERE IS LIFE ON MARS

astronomers, Mars was at a distance of 39,800,000 miles. The Expedition took 20,000 photographs of the planet. Some of these confirmed the existence of vegetation and the reality of the canals. It was found that these do not meander, as would natural water channels, streams or rivers. One indeed was observed to run for 1,500 miles straight as an arrow, without bend or diversion—something that no natural formation could be expected to do.

Enough photographs of the canals were obtained to prove that, not only do they certainly exist, but that they change in intensity, something that generations of visual observers have noted previously.

Dr. Slipher was reported as saying that the canals suggest lines of vegetation along watercourses. In view of these discoveries, some astronomers are now inclined to return to the theory put forward by Percival Lowell and others, that the Martian canals are of artificial origin.

If that were so, we can only conclude that some kind of intelligent life exists there. What form it would take is anybody's guess. Certainly, I think, nothing like ourselves, owing to the very different conditions under which it has evolved. The creatures, if they exist at all, would be bipeds, having like us the *pentadactyl* limb, arm or leg, terminating in five digits, fingers or toes—a type of limb first evolved by creatures who were gradually turning from an aquatic to a terrestial existence. They would probably be small, not more than three or four feet in height, very active, and, like us, large-brained.

Turning to the canals, another curious point, noted by the American Expedition, was that they are often observed to run straight through each other, something no natural waterway would be likely to do.

New Area of Vegetation

The Expedition discovered a vast new blue-green area, which did not previously exist. The area was estimated to be

NOTES

some 200,000 square miles, and this discovery constitutes the greatest change in Martian geography since the first maps of the planet were made 125 years ago.* This new area of Martian vegetation is in the vicinity of the Great Thoth Canal. Previously dark areas have only appeared as extensions of existing areas, and this discovery has shown that the division between the reddish orange, supposed desert areas, and the dark areas, is neither fixed nor permanent. The one may change to the other from time to time.

U. A. T.

The project for launching a small artificial satellite, or miniature space station, that will orbit the Earth at a distance of some 200 to 500 miles from the surface, which is now being pushed forward with all speed by the governments of the United States, Russia, and Great Britain, is a first step towards the ultimate goal of inter-planetary travel and the exploration of space.

These first satellites or space stations, will be quite small objects, packed with recording instruments. The information obtained by these will probably be transmitted to Earth by short wave radio. The satellites will be carried up to a height of about 200 miles or more by rocket, and then launched into an orbit round the earth at a speed of about 5 miles per second, or 18,000 miles an hour. At this height, although it is very rarefied, there is still a sufficient density of atmosphere to offer resistance to a fast-moving object, so that the satellite would be gradually slowed down and thus pursue a spiralling course towards the Earth, although it would not become a possible danger to life by crashing. Once it entered the dense lower strata of the atmosphere its speed would cause it to become heated to incandescence and it would be vaporized before reaching the surface.

* For comparison the area of the state of Texas is approximately 266,000 square miles, while the total area of the British Isles is only about 120,000 square miles.

THERE IS LIFE ON MARS

In order to remain in a permanent orbit round the Earth, space stations would have to attain a distance outwards from the Earth of 1,000 miles or more in order to be entirely clear of the atmosphere.

Incidentally, there is no such thing in space as *up* and *down*, but merely movement inwards towards a body in space, a planet or star (Sun), or movement outwards away from it. We talk about looking *up* at the stars, when actually we are doing nothing of the sort, but merely looking out at them from the Earth's surface.

All this is a first step towards space exploration, but there is a more immediate goal in sight. I have called it, for want of a better name, U.A.T., or Upper Atmosphere Travel. This will be accomplished by means of passenger-carrying vessels which will be propelled by rocket motors, or later, perhaps, by atomic power, or some entirely new type of propellant as yet unrevealed. They will take off from London, say, and land their passengers in New York within the hour. Most of the time would be taken up by gradual acceleration after the take-off, and an equally gradual deceleration prior to landing. The vessels would attain a maximum height of perhaps 250 miles, and a maximum speed of over 5,000 miles per hour.

Until now, the greatest height reached by a rocket has been 250 miles, and the highest speed attained 5,000 miles an hour, but these achievements will shortly be eclipsed.

There is of course a more sinister reason behind the frantic international race to be first with the launching of an artificial satellite.

In the future, as in the past, every new invention that can be adapted for war purposes will be so adapted at the earliest opportunity.

The first nation to launch a space station having a crew aboard, will possess a weapon of incalculable value. Armed with telescopic cameras and other instruments, the crew will be able to observe military dispositions anywhere in the world. They would be able to direct, if not control, the launching of

NOTES

nuclear weapons on any city or other military objective with the most complete accuracy.

Certainly the crew members will have other tasks to carry out. They will be kept busy observing the Moon and stars, to say nothing of Mars and the other planets. They will be obtaining data on many subjects concerning ultra-violet rays, cosmic radiation, vast electric currents circulating in the upper atmosphere, air movements there which affect the weather lower down, and much else besides. But of one thing you may be quite sure, their principal and most important task will be watching the Earth, watching other countries. If the Americans are the first up there, they will be watching, above everything else, the Russians. So shall we, for that matter.

It would be very foolish to allow ourselves to be lulled into a sense of false security by any talk of friendly rivalry, or exchange of information between the nations. Some of the world's cleverest scientists are now working feverishly on this problem of the conquest of space. It has become a desperate race between the leading countries to see who will get there first.

Whoever first really conquers space will dominate the Earth.

Forms of Life

When assessing the possibility of higher forms of life occurring on other Planets, it is a mistake, I think, to assume that life must always and everywhere follow the pattern it happens to have followed here, or that conditions similar to those prevailing here must always be essential to the genesis and evolution of life.

It may be, that, given very different conditions, life may assume forms very different to anything within the limits of our experience.

If that is so, higher forms of life, and even intelligent life might exist on planets differing so greatly from the Earth as Mars or Venus.

THERE IS LIFE ON MARS

With regard to climate, the lowest temperature ever known here was recorded at Verhoinsk, Siberia, on 3rd April 1956, 102° F. below Zero. The previous record low temperature was 90° F. below Zero, also at Verhoinsk, in 1892. The surface temperature of Mars sometimes falls in places, to 95° F. below Zero.

Bibliography

Bibliography

Astronomy and General

ANTONIADI, E. M. *La Planete Mars.* (Paris, 1930.)
CLARKE, A. C. *The Exploration of Space.* Temple Press (London, 1951).
HOYLE, FRED. *The Nature of the Universe.* Basil Blackwell (Oxford, 1950).
—— *Frontiers of Astronomy.* Heinemann (London, 1955).
JONES, SIR HAROLD SPENCER. *Life on Other Worlds.* English Universities Press (London, 1952).
LEY, W. *The Conquest of Space.* Sidgwick & Jackson (London, 1950).
—— *Rockets, Missiles and Space Travel.* Chapman & Hall (London, 1951).
LOWELL, P. *Mars and Its Canals.* The Macmillan Company (New York, 1906).
—— *Mars as the Abode of Life.* The Macmillan Company (New York, 1908).
—— *The Evolution of Worlds.* The Macmillan Company (New York, 1909).
MAUNDER, E. W. *Are the Planets Inhabited?* Harper & Brothers (New York, 1913).
MICHIELS, J. L. *Atomic Energy.* C. A. Watts (London, 1951).
NELSON, THE EARL. *Life and the Universe.* Staples Press (London, 1953).
PHILLIPS, T. E. R., and STEAVENSON, W. H. *Splendour of the Heavens.* Hutchinson (London, 1923).
STRUGHOLD, H. *The Green and Red Planet.* Sidgwick & Jackson (London, 1954).
UREY, H. C. *The Planets, Their Origin and Development.* Yale University Press (New Haven, 1952).

VACOULERS, G. DE. *The Problem of Mars:* translated by T. A. Moore. Faber & Faber (London, 1950).
WHIPPLE, F. L. *Earth, Moon and Planets.* Churchill (London, 1947).

Biology and Aviation Medicine

ARMSTRONG, H. G. *Principles and Practice of Aviation Medicine.* Williams & Wilkins (Baltimore, 1952).
DILL, D. B. *Life, Heat and Altitude.* Harvard University Press (Cambridge, 1938).
HARROW, B. *Biochemistry.* W. B. Saunders (Philadelphia, 1946).
MARBARGER, J. P. (Ed.) *Space Medicine.* University of Illinois Press (Urbana, 1951).
MCDOUGALL, W. B. *Plant Ecology.* Lea & Febiger (Philadelphia, 1949).

Space, Rockets and Space Travel

BACON, J. S. D. *The Chemistry of Life.* C. A. Watts (London, 1947).
CLARKE, A. C. *Interplanetary Flight.* Temple Press (London, 1950).
CLEATOR, P. E. *Rockets through Space.* George Allen & Unwin (London, 1936).
DAVIDSON, MARTIN. *From Atoms to Stars.* Hutchinson (3rd edition, London, 1952).
—— *Astronomy for Everyman.* J. M. Dent (London, 1953).
GIRVAN, WAVENEY. *Flying Saucers and Commonsense.* Frederick Muller (London, 1955).
GODDARD, DR. R. H. *Rockets.* American Rocket Society (New York, 1946).
HABER, H. *Man in Space.* Sidgwick & Jackson (London).
HARPER, H. *Dawn of the Space Age.* Sampson, Low, Marston (London, 1947).

BIBLIOGRAPHY

JEANS, SIR JAMES. *Through Space and Time.* Cambridge University Press (Cambridge, 1949).
—— *The Mysterious Universe.* Cambridge University Press (Cambridge, 1948).
JONES, SIR HAROLD SPENCER. *Worlds Without End.* English Universities Press (London, 1935).
—— *General Astronomy.* Edward Arnold (London, 1951).
LEONARD, J. N. *Flight into Space.* Sidgwick & Jackson (London).
NEVILLE, G. T. *Evolution in Outline.* C. A. Watts (London, 1951).
PENDRAY, G. E. *The Coming Age of Rocket Power.* Harper & Brothers (New York, 1947).
RIDLEY, G. N. *Man: The Verdict of Science.* C. A. Watts (London, 1946).
RYAN, C. (Ed.). *Across the Space Frontier.* Sidgwick & Jackson (London).
—— *Man on the Moon.* Sidgwick & Jackson (London).

Index

Index

Aeroplanes, 109, 125
Air, 114
Algæ, 30, 34
Amino-acids, 22, 37, 38
Amœba, 42
Anabolism, 26
Anærobic bacteria, 38
Anærobic respiration, 33
Andromeda nebula, 14
Antoniadi, 89, 90
Argon, 97
Arizona, 91
Artificial satellites, 110, 111
Asteroids, 13, 93
Atmospheric pressure, 32
Atom, 19, 21, 30
Atomic bomb, 21
Aurora Borealis, 57

Bacteria, 29, 30, 34, 36, 38, 43, 45
Balloons, 109
Barnard, Dr., 91
Belopolsky, 95
Bibliography, 135–137
Bio-chemistry, 40–45
Birds, 32
Boron, 16
Butterfly, 87

Cabbage, 87
Calcium, 26

California, 58, 91
Canada, 50
Carbohydrates, 23, 26, 34, 39
Carbonic acid, 49
Carbon, 26, 103
Carbon atom, 15, 17, 104
Carbon dioxide, 27, 33, 36, 39, 46, 48, 49, 50, 55, 56
Carnivores, 39
Catabolism, 27
Cells, living, 42
Chemistry, 41
Chemistry, Bio-, 40–45
Chlorine, 26, 39
Chlorophyll, 22, 23, 26, 39
Churchill, Sir Winston, 72
Clouds, Venusian, 107
Coal, 47, 52
Combustion, 27, 28
Comets, 93
Continuous creation of matter, 19
Copernicus, crater of, 102
Cosmic energy, 20, 106
Cosmic rays, 21, 110, 116

Day, Martian, 79
Death, 40, 41, 45
Deserts, Martian,
Dogfish, 87

INDEX

Earth, 13, 18, 20, 30, 46, 47, 80, 93; earth temperature, 47, 52; earth, age of, 47; earth atmosphere, 54, 57, 81, 88
Ecology, planetary, 29–36
Electron, 16
Elements, 17, 104
Energy, 19, 20; energy, chemical, 27; energy, kinetic, 27; energy, electrical, 40
Enzymes, 37
Eratosthenes, crater of, 101
Everest, Mount, 70
Evolution, 97, 105

Fats, 26
Fermentation, 27
Fission, 20
Flagstaff, Observatory, 73, 84, 90
Flying saucers, 71
Fretum, 89
Fusion, 20

Galaxy, 13
Genes, 37
Gills, 28
Glucose, 23
Gravity, 110, 111
Greenland, 51
Green plants, 22, 23, 26, 39, 49, 55

Hæmoglobin, 26
Helium, 20
Hydrogen, 17, 19, 20, 26; hydrogen bomb, 21
Hydroponic, cultivation, 124

Ice ages, 46
Infra-red radiation, 110
Insects, 79
Inter-planetary travel, 109–120
Interstellar gas, 13, 106
Iodine, 26
Ionised layers, 119
Iron, 17, 26

Kelvin scale, 30
Krakatoa, volcano, 48, 62, 64
Kuiper, G. P., 66, 78

Lacus, 89
Lichens, 29, 30, 33, 34, 106
Life, 18, 19, 21, 29, 31, 40, 41, 44, 78, 79, 104; life, evolution of, 21; life on other worlds, 103–108; life substance, 26; Martian life, 35; terrestial life, 37; primitive life, 37; microbial life, 40; microscopic life, 42, 87
Light, speed of, 119
Light year, 14
Living matter, 16, 22, 104
Lockyer, Sir Norman, 95
Lowell, Percival, 67, 84, 85, 89, 90, 91, 106
Lunar Base, 121–129
Lunar mists, 122
Lungs, 28, 70

Magnesium, 17, 26
Magnetism, 16
Mammals, 31, 32
Man, 87

INDEX

Mare, 89
Mare Crisium, 101
Mars, 18, 30, 56, 58, 92; atmosphere of, 29, 30, 32, 33, 35, 46, 57, 68, 78, 81, 89; climate of, 31, 59, 82, 83; life on, 35, 66, 70-73, 77-79, 103-108, 128, 129; canals, 58, 62, 67, 68, 70, 84-92; diameter of, 58; polar caps, 59, 60, 77, 78; deserts, 59; dark areas, 59; mountains, 59; colour changes on, 60-67; vegetation on, 60-67, 86; volcanic theory, 62-65; yellow mists on, 63, 64; clouds on, 68; satellites, 69-70; summary of conditions on, 74-76; oases, 90; landing on, 121-129
Martian base, 121-129
Martians, 71, 72, 105
Matter, 20
McLaughlin, D. B., 62
Mercury, planet, 18, 93, 94
Metabolism, 23, 27, 35
Meteorites, 117
Meteors, 56, 116, 117
Milky Way, 13
Molecule, 17, 30
Moon, 56, 58, 91, 93, 99, 100, 101, 102; landing on, 121-129; water on, 122; atmosphere of, 101, 122; life on, 100, 101
Morse signalling, 119
Mosses, 30, 34
Mount Palomar, telescope, 14, 18; Observatory, 92

Mount Wilson Observatory, 73, 91
Multiple stars, 104

Notes, 129, 131-134
Nitrogen, 16, 26, 97

Oil, 52
Öpik, Dr. E., 66, 106
Oxygen, 17, 18, 22, 23, 26, 27, 31, 32, 55, 97
Ozone, 25, 38
Oxygen pressure, 32
Oxidation, biological, 32
Oxygen, free, 35, 36, 37, 66

Pacific Ocean, 99, 100; basin of, 99, 100
Palomar Observatory, 92
Phosphorus, 26
Photography, astronomical, 91, 92
Photosynthesis, 23, 33, 35, 38, 39
Piton, Mount, 102
Planets, 18
Plants, 35
Plass, Gilbert, 48
Pluto, Planet, 94
Poles, North and South, 52
Potassium, 26
Proteins, 22, 37
Protons, 116
Protoplasm, 26, 39
Protozoon, fresh water, 23

151

INDEX

Radiation, 30
Radio-activity, 16
Radio, 110, 119
Rainfall, 50
Relativity, 15
Reptiles, 31
Respiration, 27
Rockets, 110
Rocket propulsion, 109–112
Rockets, to Moon, 111
Russia, 51
Rust, 55

Schiaparelli, 84, 89, 91
Sea, 51
Siberia, 31, 50, 51
Silicon, 17, 26, 103
Sinus, 89
Sodium, 26
Solar system, 13
Solar energy, 23, 103
Space navigation, 118
Space ships, 109–120
Space stations, 110, 111
Spectroscope, 95
Spitzbergen, 51
Stars, 104–106
Stevens and Anderson, 116
Sulphur, 26
Sun, 20, 21, 30, 80, 93, 94
Superior conjunction, 93
Switzerland, 51
Symbiosis, 34

Time, 20
Trachea, 28
Track vehicles, 125
Tripoli, 31
Troposphere, 88
Twilight, 93
Tycho, crater of, 102
Tyndall, John, 48

Ultra-violet light, 25
Universe, 15, 17, 18, 40, 87, 97

Vaucouleurs, G. de, 90
Vacuum, 109
Velocity, 118
Velocity of escape, 53, 54, 56
Venus, 18, 30, 46, 56, 71, 93–102; life on, 71, 87, 88, 96, 97, 103–108; transit of, 94; atmosphere of, 95, 96, 98, 99, 107, 108
Venusian year, 94, 95
Venusian on Earth, 71, 72
Virus, 44
Volcanic activity, 48
Volcanic ash, 123

Water, 33, 85
Wright Brothers, 109

152